Predation and Freshwater Communities

Predation and Freshwater Communities

Thomas M. Zaret

Foreword by G. Evelyn Hutchinson

New Haven and London
Yale University Press

Published with assistance from the foundation
established in memory of Philip Hamilton McMillan
of the Class of 1894, Yale College

Designed by James J. Johnson
and set in Optima type.
Printed in the United States of America by
Edwards Brothers Inc., Ann Arbor, Mich.

Library of Congress Cataloging in Publication Data

Zaret, Thomas M 1945–
Predation and freshwater communities.

Bibliography: p.
Includes index.
1. Freshwater ecology. 2. Predation (Biology)
3. Competition (Biology) 4. Natural selection.
I. Title.
QH541.5.F7Z37 591.5′2632 80–5399
ISBN 0–300–02349–9

10 9 8 7 6 5 4 3 2 1

Contents

Illustrations

Tables

Foreword

Fifteen years ago, no data in limnology made less sense than those relating to the specific composition of the zooplankton faunas of different lakes. In 1965 a classical paper by Brooks and Dodson showed the way to resolve much of the difficulty of the subject. Since then a flood of light has been thrown on what had been exceedingly obscure. In this illumination, no one has been more active than the author of this book. His wide field experience, extensive learning, and sound judgement in the development of a simple theoretical approach not only open up to the limnologist what has seemed to be an intractable field but more importantly bring into the mainstream of contemporary ecological biology a large amount of interesting and conceptually significant material. I am very happy to be, even in a small way, associated with its publication.

G. Evelyn Hutchinson

Preface

Charles Darwin wrote in *On the Origin of Species* (1859) that when the process of natural selection resulted in species extinction, it was due exclusively to one species outcompeting another (p. 110). Succeeding generations of ecologists, from either Darwin's direct influence or because they shared the same biases in interpreting the natural world, have relied primarily on the grindstone of competition theory to explain everything from species morphological complexity to the behavioral repertoires of animals. Competition was accepted as the dominant biotic process, the fundamental basis for ecological theory, and the driving force of evolution. Small wonder, then, that the contrasting explanation for the causes of extinction developed by Alfred Russel Wallace remained relatively unappreciated among ecologists until a proponent of predation phenomena pointed out the poignant differences (Brooks 1972). For Wallace, predation was one of *several* selective forces capable of driving species to extinction; his mechanism was not restricted to competitive interactions as was Darwin's. Arguments that consider whether it was Darwin or Wallace who developed the first theory of natural selection are therefore more than mere discussions among historians of science for the assignment of priority of authorship. Regarding the role of competition and predation in natural communities, the arguments concern the question of which interpretation of the theory, Darwin's or Wallace's, best conforms to our current knowledge of the evolutionary process.

To examine this argument we may ask: "What is the relative importance of competition and predation in determining the organization and structure of natural communities?" The strength of the competition paradigm is witnessed in the literally thousands of scientific articles that have provided support for the importance of competition. The majority of these assume a priori that competition is the dominant force, develop theory based on this assumption, and marshal facts from field studies that support the competition point of view. The weakness of this single-minded approach is that in many cases results can be predicted for other reasons as well; in addition, alternative explanations are not always considered.

In this book I try a different approach to examine the relationship between competition and predation. I first develop a series of models based on the paradigm that predation processes can explain the structure of natural communities. Predation is thus used artificially as the driving force. From inductive reasoning, community models are developed with zooplankton communities as an example. Then predictions are generated and tested with field data. Where predictions are supported by field tests, it is assumed that predation is the dominant factor; where predictions fail to be supported, predation can be considered unimportant relative to some other factor such as competition. The objective is to identify where predation is the best predictor of community characteristics, where competition is the best predictor, and where neither applies significantly. The goal is to increase our understanding of the interaction of these two biotic processes and to further our knowledge of the evolutionary process of natural selection as first proposed by Darwin and Wallace.

Acknowledgments

The first draft of this book was begun on Barro Colorado Island in Panama in February 1974. Typing during the early morning hours, I listened, fascinated, to the dawn chorus of the black howling monkeys, shouting out from a thick, green tropical world and somehow expressing my own frustrations at trying to put ideas down on paper. Several years and several drafts later, I finished the final version in the wet Seattle October. During the agonizingly long gestation period of this book, I have had the support of numerous friends and colleagues. My thanks go to my departmental colleagues, R. T. Paine and W. T. Edmondson, without whose criticism and support, both financial and scientific, this book would never have been completed. Various important improvements and suggestions were made by Bob Black, Nelson Hairston, Jr., Charles Kerfoot, John Lehman, Barbara Taylor, and Bob Pastorok. Thanks also go to John Bizer and Owen Sexton for their help on interpreting salamander life histories, and to Richard Wasserzug, who evaluated some of my interpretations on anuran filtering abilities. I wish to thank especially Al Covich, Mike Lynch, Bill Neill, and Mike Swift for their helpful criticisms after reading earlier drafts, and Doug Eggers for discussions on planktivore foraging models. A host of friends who provided technical support include Griselda Alsina Brandani, Bonnie Brewer, Carol Casinelli, Bruce Hanson, Bob Jackman, Joyce McKenney, Barbara Peterson, and Carol Slocum. My greatest appreciation is reserved for my wife, Kim, who has served every role from scientific consultant, data gatherer, editorial critic, shoulder to lean on, and contributing illustrator for most of the book's figures. Without her support I would never have been able to write this book. To these and other persons too numerous to mention, I dedicate this book.

Most of my original ideas seem to depend on the rigors of field work for their final transition from half-developed notions to plausible hypotheses. Since 1967 I have spent approximately a quarter of each year engaged in studies of aquatic systems, mostly in the New World Tropics, which serve as my vital source of inspiration. I have had the opportunity to work in Mexico, especially in the high plateau lake region between Mexico City and Guadalajara; Guatemala, including Lakes Amatitlán and Atitlán and the fault lakes of the Petén; El Salvador; Honduras; the Great Lakes of Nicaragua; Costa Rica; the Gatun Lake–Chagres River Basin of Panama; the valley lakes of Colombia; Ecuador; Lake Titicaca in Peru and Bolivia; throughout Venezuela from the lowland llanos to the very top of Tepuy Roraima; and most recently in the Amazon Basin of Brazil. During these many expeditions I have worked with and been supported by local scientists and citizens and I express my deepest gratitude to those numerous and friendly people who have made my work possible.

The research upon which this book is based has been generously funded (in chronological order) by Yale University, Connecticut; the Smithsonian Tropical Research Institute, Balboa, Panama; the Organization for Tropical Studies; and the National Science Foundation. I am extremely grateful for the support and generosity of these agencies and also to the taxpayers of the United States for their support of basic research through these institutions.

Finally, I wish to acknowledge two scientists, Dr. John L. Brooks and Dr. Charles Elton, whom I feel privileged to know and who have served as a special inspiration to me during my work.

1. Introduction

This book examines the role of predation in natural communities, using as an example lacustrine systems. As a focal point for the considerable literature on these systems, a series of biological models have been developed, which I term "species-interactive models," in order to explore the structure and organization of freshwater communities. The major portion of the book concerns the development of these models, beginning with a detailed exposition of their empirical bases (chapters 1–5), a synthesis of the various components (chapter 6), and finally a series of tests to assess their explanatory power (chapter 7). The final two chapters draw on the model results to consider both competitive effects (chapter 8) and finally their applicability to other systems and the implications for ecological theory (chapter 9).

I use the physical environment and its effects on the community only superficially, but this does not mean that I consider such physical limitations of minor importance. On the contrary, unless the species meets the rigors imposed by the physical environment, it will fail to persist (see the considerable literature devoted to investigation of the physical factors of lakes; e.g., Hutchinson 1967). For these species-interactive models, however, one can assume the obvious, namely, that the animals associated with a given lake are able to survive any manifestations of the physical environment that occur normally. Considerations of the genetic constitution and phenotypic plasticity of species are important but not necessarily critical for making ecological predictions about the community models. Thus, we would not ask why there are no coregonids (whitefish) in the southwestern United States. Rather, given the piscine planktivores in that region, we would ask why in a particular lake we find one species of fish rather than coregonids, and what the resulting effect is on the rest of the community.

In general, when we find one kind of lake, we will find several lakes of the same kind, for whatever natural force created the first lake basin usually will have acted over a wide geographical area and be responsible for many such basins in the same region. If we find one tectonically produced lake (i.e., resulting from deformation of the earth's crust), such as a fault lake, we will usually find others along the same fault—for example, the Great Lakes of Africa or the series of fault lakes in Petén, Guatemala. A kettle lake in the northeastern United States, the product of recent Pleistocene glacial action, will be associated with hundreds of other kettle lakes. One volcanic core lake in Central America will indicate others nearby. This phenomenon has given rise to the concept of "lake types" (Hutchinson 1967). Once we know the origin of the lakes in an area, we can assume that there will be a certain similarity among them. If we are examining a region where the lakes are formed by wind-associated processes, and thus are shallow, we can assume that zooplankton vertical migration patterns will not be very important. If we are considering a region of high altitude or alpine lakes, we can expect an absence of fish because these climates can be inhospitable or physically inaccessible, and we can expect, instead, other zooplankton predators. In lakes heavily influenced by river waters, we can anticipate a zooplankton community characteristic of lotic (rapidly moving) conditions and dominated by littoral (nearshore) or benthic (bottom-associated) organisms. These are the ways in which knowledge of the physical environment aids in developing predictions for the models.

In the Appendix, "Further Considerations," various thought-provoking questions pertaining to each chapter are presented. Our current understanding would be greatly enhanced by investigation of many of these questions, which should be interesting to the reader whose appetite for ecology has just been whetted as well as to the reader whose knowledge includes most of the contents of this book.

2. Class I: Gape-Limited Predators

The dominant predators found in freshwater communities are termed "gape-limited predators" because the predator's mouth diameter, or gape, determines the maximum size of prey it can take. Because this type of predator swallows its prey whole, once the size of the prey exceeds the diameter of the predator's mouth, the probability of its being eaten is zero. Gape-limited predators (GLP) are defined as predators exhibiting a prey electivity* curve that increases with some aspect of prey attractability such as body-size. In other words, this is a probability function: the larger the prey, the more likely it will be taken when available by the predator (fig. 1). The following section discusses in some detail several groups of GLP, their characteristic prey selection, and the resultant effects on zooplankton communities.

SELECTIVE FEEDING BY FISHES

Among planktivores (zooplankton predators), freshwater fishes are the best understood and most thoroughly studied, having received special attention during the last decade, with considerable effort focused on their role in lake communities. That the electivity of piscine planktivores for a particular prey item can be predicted from that prey's body-size was first illus-

* In this discussion I use Ivlev's definition of electivity (1961), which is simply the ratio of what is eaten by the predator to what is actually present in the water column. This differs from "selectivity," which refers to the predator's choice given many objects equally available. These terms will be discussed in more detail later.

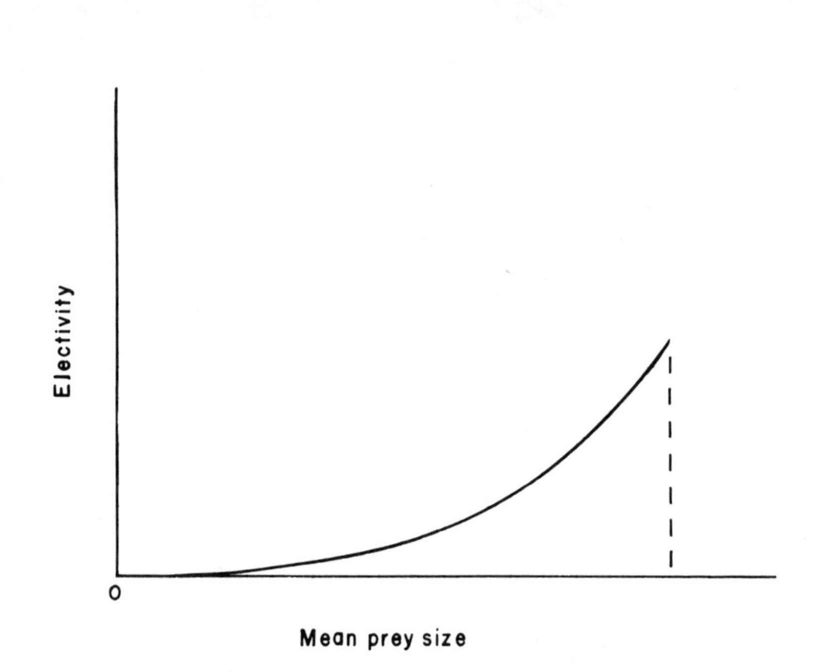

Fig. 1. Electivity curve for gape-limited predators

trated in a laboratory study by Brooks (1968), who presented different sizes of the calanoid copepod *Diaptomus minutus* to captive alewives (*Alosa pseudoharengus*, Clupeidae). In a large aquarium containing the experimental fish, Brooks introduced plankton samples that included five copepod size classes: the copepod larval stages known as nauplii (mean length 0.10 mm); the first juvenile, or "copepodid," stage (mean length 0.39 mm); the second copepodid stage (mean length 0.49 mm); the third copepodid stage (mean length 0.60 mm); and adult copepods (mean length 0.76 mm). The experiment lasted one hour. After every fifteen minutes of fish feeding, a subsample of the water was evaluated to determine the remaining abundances of the different-sized prey categories. After the first fifteen minutes virtually none of the smallest nauplii category had been consumed by the fishes, 10 percent to 55 percent of the copepodid stages had been eaten, and 65 percent of the adult *Diaptomus*. This preferential removal of the largest adult-size classes continued throughout the feeding experiment until, as indicated in figure 2, the end of the hour when only the smallest nauplii remained in the aquarium. A comparable laboratory feeding experiment was performed by Werner and Hall (1974), who used juvenile

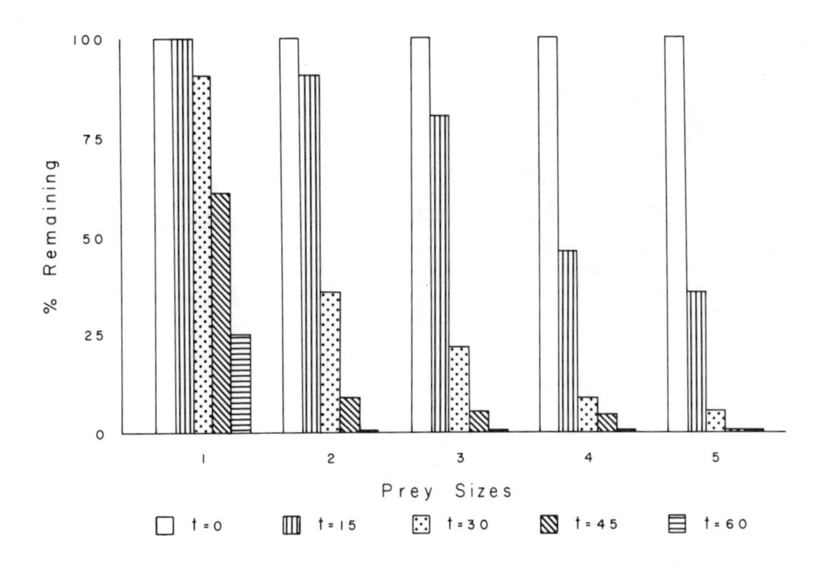

Fig. 2. Effect of prey size on predator selectivity. *Alosa pseudoharengus* feeding on *Diaptomus minutus*. Ordinate gives percent of prey remaining at five different time periods (*t*) of the one-hour experiment ($t = 0$ is start, 15, 30, 45, and 60 minutes) for each prey size category on the abscissa (1 is nauplii, mean length = 0.10 mm; 2 is copepodid = 0.39 mm; 3 is copepodid = 0.49 mm, 4 is copepodid = 0.60 mm; 5 is adult = 0.76 mm). (From Brooks 1968)

bluegills, *Lepomis macrochirus* (Centrarchidae), feeding on different size categories of the cladoceran *Daphnia magna* (fig. 3). In four experiments, the small bluegills (70–80 mm total length) were presented with four size categories of daphnids: 3.6 mm, 2.5 mm, 1.9 mm, and 1.4 mm. Half the experiments had twenty-five individuals per prey size, and half had fifty individuals. For the four experiments, the means of feeding ratios for all prey size classes were 1 : 0.83 : 0.54 : 0.27, demonstrating that, given equal numbers of prey classes, planktivore electivity is proportional to prey body-size. These data form the basis for the predator-typical curve in figure 1. Only after the preferred large size categories decline in abundance do the fish begin significant feeding on the smaller size categories.

Although planktivores feed preferentially on the largest prey items available in their environment, I have categorized fishes as "gape-limited predators" because of one possible exception. For the youngest stages of fish populations (i.e., the larvae) there may be a brief time when the small

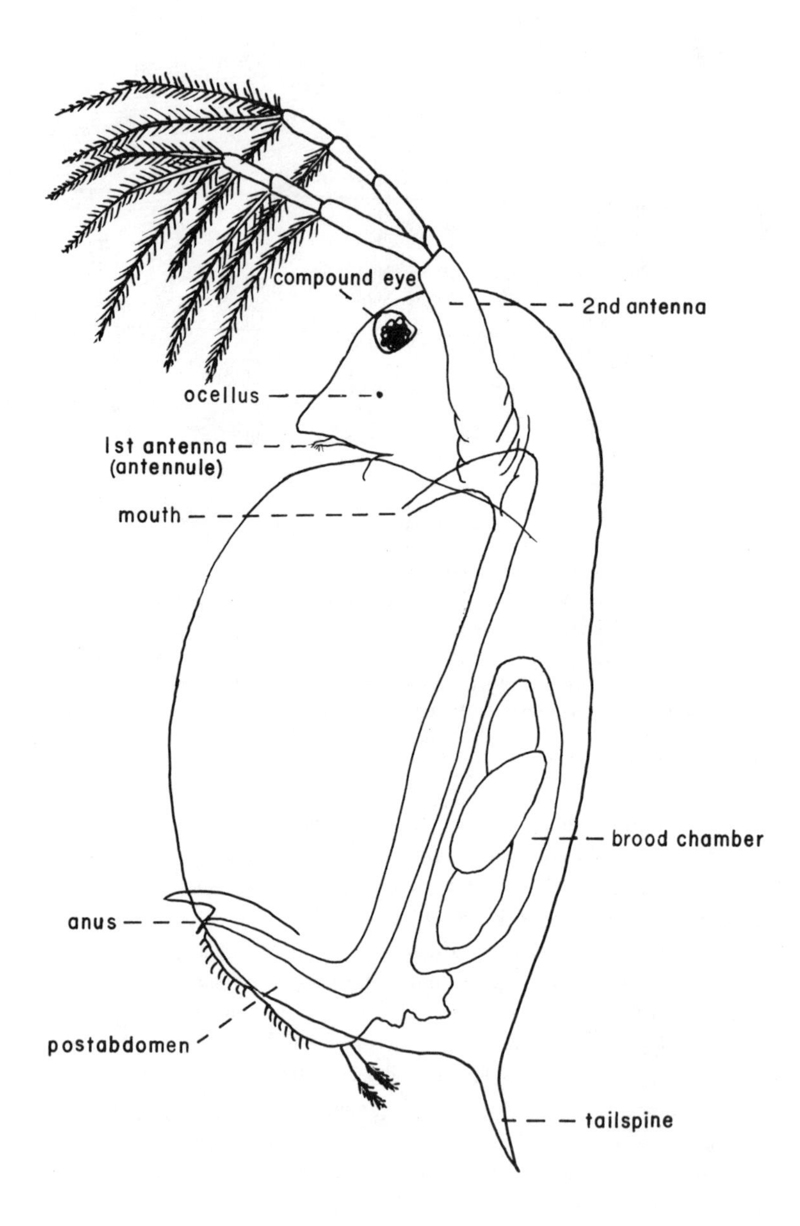

Fig. 3. Simplified diagram of *Daphnia* (Crustacea: Cladocera)

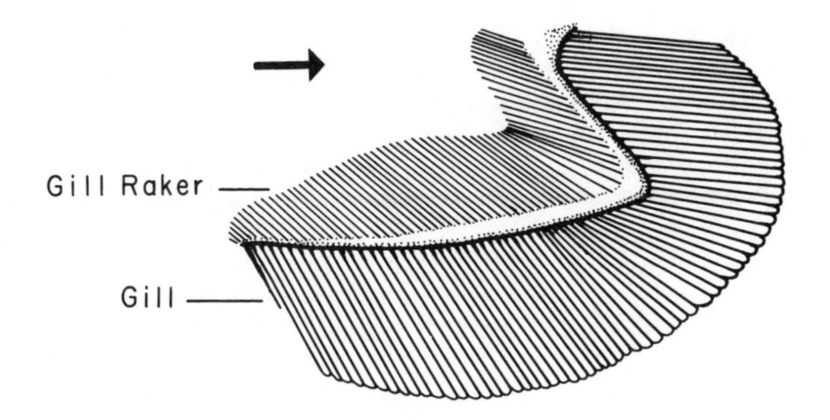

Fig. 4. Planktivore branchial arch with gills and gill rakers as indicated. Direction of waterflow indicated by arrow.

size of their mouth, or gape, prevents them from consuming the largest prey in their environment. As the larvae grow, however, and their gape increases, they rapidly become more efficient at prey capture until they can consume the largest items normally found in lakes (Rosenthal and Hempel 1970; Wong and Ward 1972). For instance, Werner and Hall (1974) have found that by the time the bluegill attains a size of 50 mm, it can easily handle the entire range of benthic insect prey available. Limnetic prey are of a smaller size range. Values for marine fishes indicate that larvae attain a 90 percent feeding success within a few days of hatching and are able to feed on the maximum available size range of prey by three weeks of age (Hunter 1979). Thus, this gape limitation is of only very brief duration in the fish's life and can be ignored for most considerations of freshwater community interaction.

The gill rakers, cartilaginous structures built like fine combs on the gill arches of planktivores, help collect prey items from the water column (fig. 4). Because of the great numbers of planktivore gill-rakers, many people cling to the fallacious notion that all these fishes are filter-feeders. To the contrary, a key to understanding the role of planktivores is the realization that these fish are *not* simply indiscriminate filter-feeders moving through the water with their mouths open like finned sieves. Rather, studies of freshwater planktivores indicate that they are highly discriminating particle selectors, whose efficiency enables them to swallow several thousand individual prey items in just a few hours. Filter feeding is rarely employed.

This has been demonstrated most conclusively by Seghers (1974) in his laboratory feeding experiments with *Coregonus clupeaformis*, the lake whitefish (Salmonidae). By observing and recording the frequency of feeding "nips" by the fish, Seghers calculated a maximum rate of 2,400 nips per hour. Examination of the stomach contents of his experimental animals showed that some *Coregonus* were ingesting more than 600 *Daphnia magna* in just fifteen minutes, corresponding precisely to the observed frequency of nips. This study indicates that the presence of thousands of zooplankton individuals in *Coregonus* stomachs is possible through particle selection. (Perhaps the first observation on individual prey selection characteristics of *Coregonus* was noted in Bohmann et al. 1940.) In another study, Galbraith (1967) undertook an investigation of yellow perch, *Perca flavescens* (Percidae), and rainbow trout, *Salmo gairdneri* (Salmonidae), which normally feed on *Daphnia* in several Michigan lakes. By carefully measuring the spacings between the proximal ends of the gill rakers, he found that for fish less than 30 cm long, the majority of gill-raker openings were less than 1.1 mm wide. This finding suggests that if the fish simply filtered water through the gills there would be a greater concentration of small plankton (equal to or below 1.1 mm) because the smaller size categories were numerically dominant in the plankton of these lakes.

In spite of this bias toward the smaller gill-raker openings, the analysis of the stomach contents of these planktivores showed a complete absence of prey less than 1.3 mm in body-length, even though the gill rakers were capable of trapping these smaller sized animals (fig. 5). Galbraith's field study clearly demonstrated that planktivore prey capture is not merely a mechanical function, such as the filtering of water, but is a highly selective adaptation by these predators. Other studies of planktivore gill rakers have supported this conclusion of particle selection. Kliewer (1970), for example, in his stomach-contents analysis study of lake whitefish populations in seven lakes of northern Manitoba, Canada, found no correlation between absolute gill-raker spacing and prey body-size.

Although filter feeding has been observed in clupeid planktivores (Blaxter and Holliday 1963; Janssen 1976, 1978), particle selection is still the dominant feeding behavior. Particle selection allows the fish to forage over a much greater area per unit time and to capture prey that would otherwise escape, in the sense that plankton, especially large swimming forms, exhibit significant detection and avoidance of plankton-capturing filter-feeders (see Fleminger and Clutter 1965). However, a more basic ev-

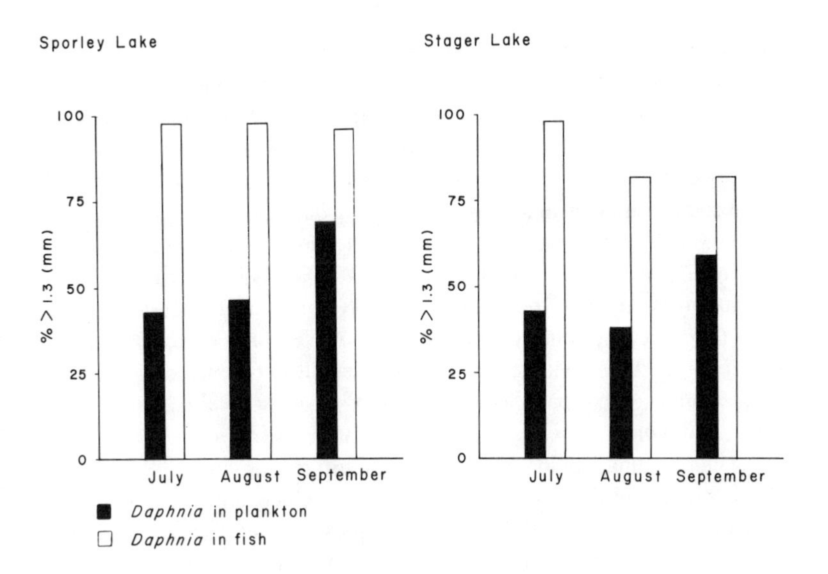

Fig. 5. Comparison from two Michigan lakes of the average percent of large *Daphnia* (>1.3 mm length) found in plankton versus that found in stomachs of *Perca flavescens* and *Salmo gairdneri* during summer. (From Galbraith 1967)

olutionary reason for the dominance of particle selections appears to be that in fresh and marine waters the density of food particles is rarely high enough to make filter feeding an effective method; that is, the density of food particles is normally so low that a filtering fish would process a great deal of water but very little food.

This conclusion is supported by Leong and O'Connell (1969) and O'Connell's (1972) laboratory studies with *Engraulis mordax*, the northern Pacific anchovy (Engraulidae), in which he varied the densities of two prey items: adult *Artemia* (anostracan crustaceans known as brine shrimp) of 3.7 mm mean body-length and 0.5 mg mean dry weight and *Artemia* nauplii of 0.65 mm mean body-length and 0.00145 mg mean dry weight. When high densities of the small nauplii were added, the fish filter-fed exclusively. When adult *Artemia* were present, however, even at low densities, the fish increased its frequency of particle-selection behavior. According to the authors, this switch to particle selection is adaptive because it accumulates biomass more rapidly than would filter feeding (Leong and

O'Connell 1969), presumably for the two previously mentioned advantages associated with particle selection. In this way, the fish can successfully exploit the adult *Artemia* when present. Furthermore, O'Connell (1972) was able to calculate from a consideration of prey biomass ingested that once the adult *Artemia* constituted 5.5–8 percent or more of the total dry weight of prey available, filter feeding would contribute little to total biomass ingestion compared to particle selection. By applying all of this information to the size distribution of prey in the fish's natural habitat in the northeast Pacific Ocean, Leong and O'Connell concluded that a reliance on particle selection would be necessary for the successful accumulation of food by *Engraulis*. Thus, given normally encountered densities of prey, it appears that particle selection is a necessity for fish to survive in their natural environment, even for fishes capable of filter feeding.

Filter feeding would be effective if the fish could remove all food particles, especially those below 0.1 mm size, which are by far the most abundant. I suggest that filter feeding in fishes is an adaptation for feeding not on zooplankton but on phytoplankton. The few fishes that consistently filter-feed are primarily phytoplankton feeders. Two strictly freshwater filter-feeders, the African *Tilapia* (Cichlidae) and the South American catfish, *Hypothalmus* (Hypothalmidae), live in environments associated with dense algal blooms and feed on phytoplankton (Greenwood 1953; Moriarity and Moriarity 1973; Roberts 1972). This indicates that the specialized filtering apparatus in Engraulidae, the hinging of the jaw, may have first evolved in phytoplankton specialists such as *Engraulis ringens*, the Peruvian anchovy (formerly the most productive fishery in the world and responsible for more than 30 percent of the total world fish production; see Johnstone 1972). Stomach-content analysis of *E. ringens* showed that 50 to 100 percent of the individuals examined contained phytoplankton, depending on the time of year, the chronologic age of individual fish, and the location of the subpopulation sampled (Rojas de Mendiola 1971). For instance, the Atlantic (*Brevortia tyrannus*) and Gulf (*B. patronus*) menhaden feed on zooplankton using particle selection when they are juveniles. (The juvenile stage is defined by size: up to 40 mm "fork" length, i.e., the length as measured to the V-shaped fork in the caudal fin.) As adults, they switch to filter feeding almost exclusively on phytoplankton (June and Carlson 1971). This shift, from particle feeding on zooplankton as juveniles to filter feeding on phytoplankton as adults, is known to occur also in the gizzard shad, *Dorosoma cepedianum* (Kutkuhn 1957). Filter feeding on zooplankton by freshwater fish is rarely found, even in laboratory studies when prey densities were increased to an order of magnitude above natural conditions (Seghers 1974).

Aberrant planktivore feeding behavior may occur where there are unusually dense concentrations of prey, either stuck to the water surface or concentrated in the surface foamlines (see Stavn 1971). McNaught and Hasler (1961) reported that the adult white bass, *Roccus chrysops* (Centrarchidae), locates unusually high densities of *Daphnia* caught in surface films and makes short swimming bursts to the surface with mouth agape, breaking the surface, to swallow the trapped daphnids. This odd behavior probably relates to the fact that adult *Roccus* are generalist feeders, normally consuming insects. In contrast, juvenile members of the population feed exclusively on zooplankton, strictly as particle selectors. As *Roccus* becomes larger and switches from zooplankton specialist to feeding generalist, it is apparently able to take advantage of the large daphnid food packets trapped on the lake surface. With one gulp, *Roccus* treats the daphnids as single prey items but actually ingests many animals at one time.

Although some fishes are capable of filter feeding, there is little evidence that this behavior is important for feeding on zooplankton in natural situations. Particle selection is the dominant behavior, and lake planktivores are highly discriminative and selective predators. In the unusual example of filtering by alewives (*Alosa*), they were found to be feeding in a nonselective manner (Janssen 1978). The major environmental cues used by fishes to detect prey are discussed in the following section.

PREY DETECTION

Dependence on Light

Because fish predators are highly selective of individual food items, it is not surprising that they are dependent on light to discriminate among prey particles. Many studies have demonstrated that the feeding efficiency of planktivores, sensitive to light, drops off rapidly as light levels decrease—the North Atlantic herring, *Clupea harengus* (Blaxter 1966); the Pacific salmon, *Oncorhynchus* spp. (Brett and Groot 1963; the bleak, *Alburnus alburnus* (Ivlev 1961); the bluegill, *Lepomis macrochirus* (Werner 1969); and the golden shiner, *Notemigonus chrysoleucas* (Suffern 1973). In his review article Hunter (1979) suggests that this generalization is applicable to all marine fish larvae; studies indicate that an absence of rods in the retina during early larval life history restricts feeding to daytime hours.

Suffern demonstrated empirically in the laboratory the correlation of fish feeding efficiency and illumination. Using an enclosed feeding chamber, he recorded the number of *Daphnia galeata mendotae* eaten by shiners (Cyprinidae) in a given time period for several different light intensities. As shown in figure 6, there was no feeding by the fish at zero light intensity, a low feeding-level threshold between 9.5 to 800 ergs cm^{-2} sec^{-1}, and a high level of feeding rate once light reached 2,200 ergs cm^{-2} sec^{-1}, corresponding to normal daytime light intensities. He showed also that size selectivity was light dependent. The importance of light for planktivore selection is supported further by studies investigating the nature of fish vision, which demonstrate the dependence of fishes on light contrast for prey detection (Hemmings 1966; Hester 1968). Additionally, Confer and Blades (1975) observed that the experimentally determined reactive distance (i.e., the distance at which the prey can be detected by the predator) of the pumpkinseed, *Lepomis gibbosus*, decreases as light intensity drops (and see Eggers 1977 for similar conclusions). Many planktivores also cease feeding after sunset when natural light conditions approach zero intensity, although bright moonlit nights can provide sufficient illumination for some feeding (Narver 1970; Blaxter 1966; Zaret and Suffern 1976).

Although piscine planktivores are usually visually dependent, there is, as usual, an exception to the rule. Members of catfish families such as the Ictaluridae, the North American catfishes (bullheads), are nonvisual feeders yet are able to consume a significant fraction of the plankton. The "barbels" of the catfish—whiskerlike extensions in the buccal region—clearly indicate the tactile feeding mode by which they consume larvae of the phantom midge, *Chaoborus* (fig. 7), as well as other zooplankton species (Costa and Cummins 1972; Pastorok 1978). Whether the catfish feeds by actively swimming at the water surface as a nocturnal planktivore or whether it feeds on benthic prey during the day by using the barbels and probing in the lake bottom may depend on the density and migration patterns of the prey. In any event, these activities are not light dependent.

Visibility

The effects of piscine planktivore feeding on the community have been examined repeatedly by workers over the last decade. One of the first important papers (Brooks and Dodson 1965), which drew upon the previous

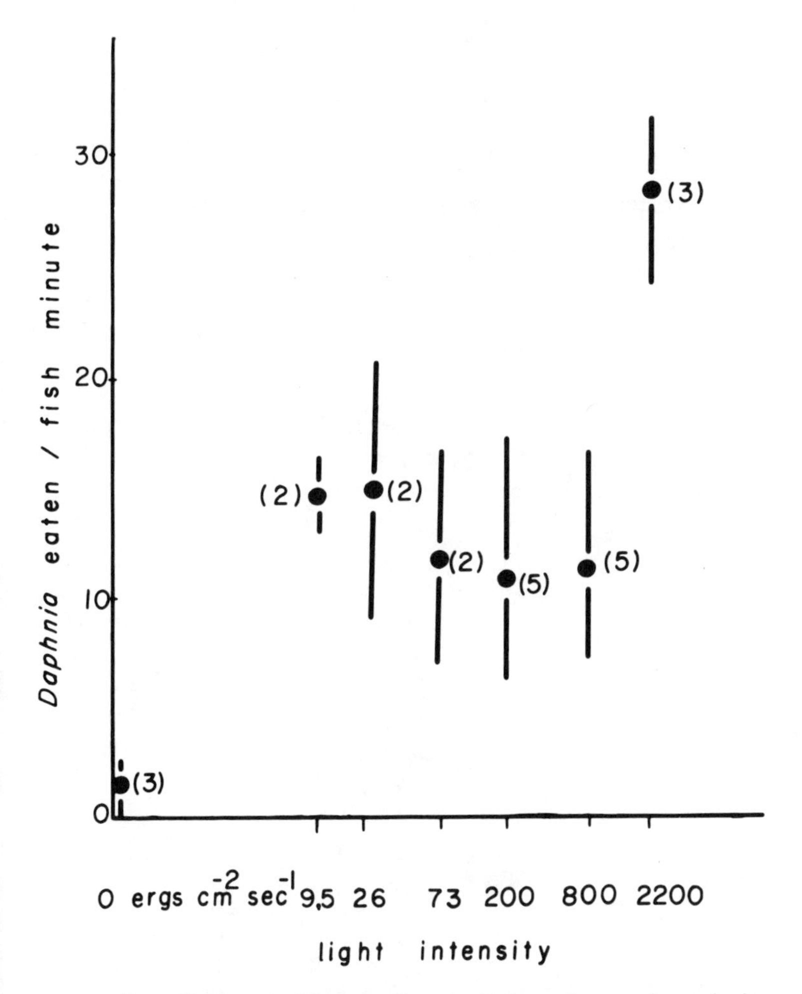

Fig. 6. Effect of light on planktivore feeding rate. *Notimegonus crysoleucas* feeding on *Daphnia galeata mendotae* at different light intensities. (From Suffern 1973)

studies of East European workers (Hrbáček et al. 1961: Hrbáček 1962), developed a unified theory relating zooplankton community changes to the actions of planktivores. (This study will be discussed in detail later.) During an examination of the zooplankton distributions of some southern New England lakes, the authors observed two distinct lake types characterized by different zooplankton communities. One lake had large zoo-

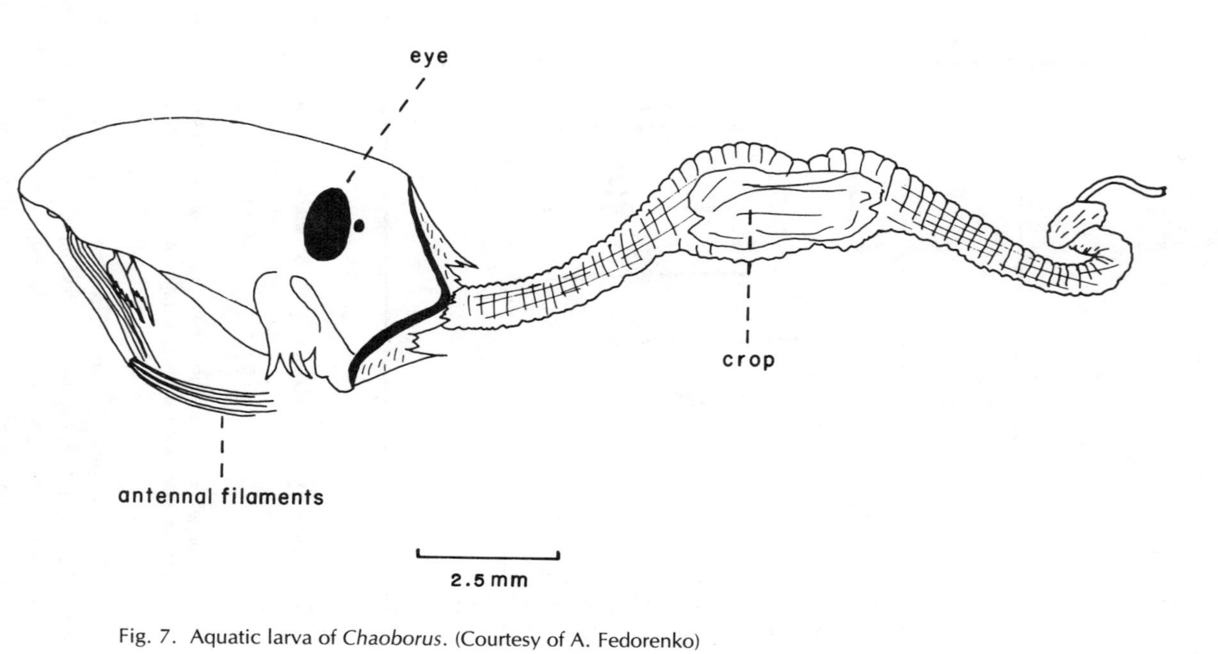

Fig. 7. Aquatic larva of *Chaoborus*. (Courtesy of A. Fedorenko)

plankton species, including members of the cladoceran genus *Daphnia*, the large calanoid copepod *Epischura*, and the cyclopoid copepod *Mesocyclops*. In contrast, several of the lakes along the eastern Connecticut coast had none of these large zooplankton species but were dominated instead by small zooplankton, including the cladoceran *Bosmina* and small cyclopoid copepods. One conspicuous difference between these two types of lake communities was the presence in the Connecticut lakes of populations of the alewife, *Alosa pseudoharengus*, a fish of marine origin that has established populations in lakes with recent marine connections.

Brooks and Dodson proposed that the differential predation by alewives resulted in the elimination of the larger lake zooplankton and their replacement by smaller forms less susceptible to predation. To test this hypothesis, the authors compared the zooplankton species found in Crystal Lake in 1942, when no alewives were present, with those species found in 1964, ten years after the introduction of a related species, the blueback herring (glut herring), *Alosa aestivalis*. They also compared and contrasted other Connecticut lakes with and without naturally occurring alewife populations.

In all cases, the presence of alewives was associated with a zooplankton assemblage dominated by the smaller-sized species (fig. 8). As further support for their argument, Brooks and Dodson (1965) referred to data from European carp ponds (Hrbáček et al. 1961), which also compared zooplankton species composition before and after the introduction of carp, showing basically the same qualitative changes. Brooks and Dodson hypothesized that it was the size-selective nature of the carp that resulted in the observed changes in zooplankton composition. The changes in zooplankton composition had other, far-reaching effects, extending to the level of the primary producers (phytoplankton).

Brooks and Dodson's publication was followed shortly by others that further documented the highly selective nature of planktivores and the size-related changes in zooplankton composition resulting from their predation. Galbraith (1967) documented the size-frequency changes in one lake's zooplankton community composition as a result of the introduction of planktivorous fishes. Ivlev's classic work on fish feeding (1961), summarizing ten years of previous Russian research, also supports the importance of size to fish predators. Other important studies include that of Green (1967), who related the differential selection of planktivores for two prey morphs to differences in prey body-size, and that of Hall, Cooper,

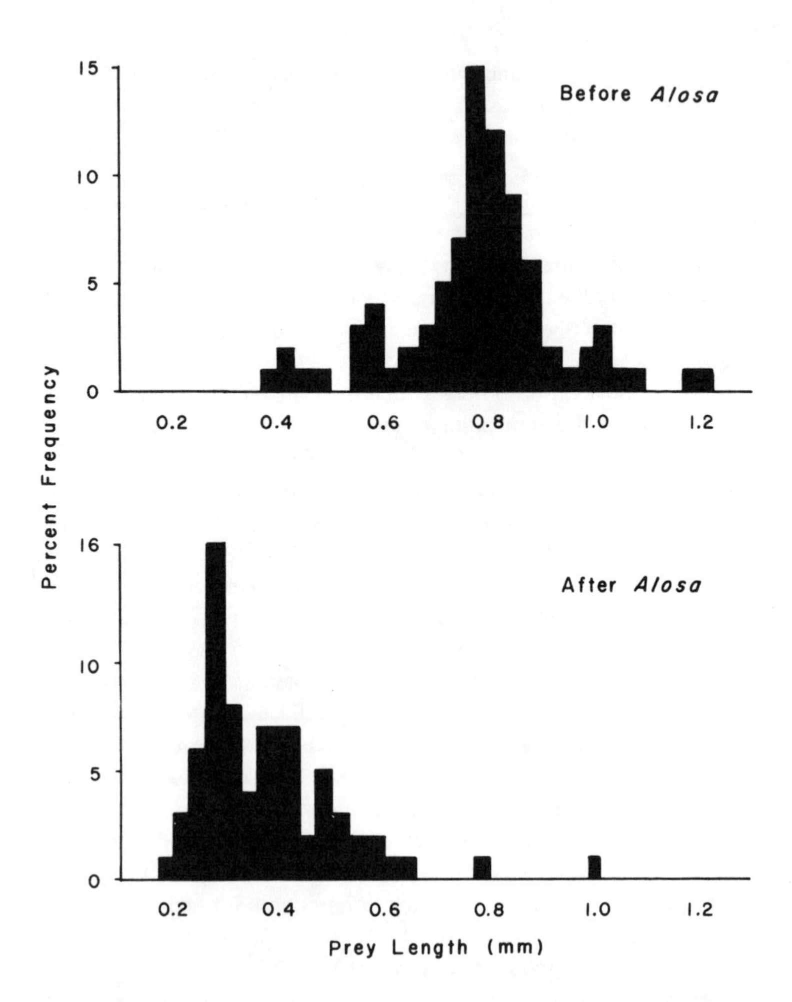

Fig. 8. Mean size of zooplankton in Crystal Lake, Connecticut, comparing ten years before the introduction of *Alosa aestivalis* with ten years after. (From Brooks and Dodson 1965)

and Werner (1970), which documented the size shifts in zooplankton populations in artificial ponds following the introduction of planktivores. Reif and Tappa's 1966 study showed replacement of a large daphid population by a smaller species following an increase in smelt abundance. Wells (1970) and Warshaw (1972), through intensive analysis of fish stomach contents or abundance correlations of natural prey, demonstrated how large zooplankton species return following the decreasing abundance of a specific planktivore, the alewife. By 1972, the idea of size-selective predation by planktivores was well on its way to acceptance. The mechanism for prey selection, however, was not understood. Did the fish really "see" the size of its prey?

In a study involving two morphs of the cladoceran *Ceriodaphnia cornuta*, with preference by the freshwater planktivore *Melaniris chagresi* (Atherinidae) for one over the other, no difference was found in prey body-size between the two morphs to explain this predator preference (Zaret 1969, 1972*a*, 1972*b*). Another possible explanation of the predator's preference was that one morph exhibited different behavior patterns—swimming motions or diel vertical migration—that were more attractive to the planktivore than the behavior patterns of the other morph. This hypothesis was also tested but found to be lacking. The only significant difference between the two morphs was that the eyes of one were up to 80 percent greater in area than the eyes of the other. Because, as discussed, planktivores are visually dependent particle selectors, relying on the contrast between an object and its background for prey detection, it was hypothesized that the predator was attracted by the large black compound eye of these otherwise transparent cladocerans (figs. 9a, 9b, and 9c).

The hypothesis was tested by a series of laboratory feeding experiments using the native planktivore, the silverside *Melaniris chagresi*. For the first experiment, the fish were fed a mixture of both large-eyed and small-eyed morphs. Substantiating field observations, the fish exhibited a greater electivity for the large-eyed form. Then the small-eyed morphs were fed a solution of water and india-ink particles, which, when ingested, resulted in a greatly increased area of black pigment just behind their eye region. This treatment did not change the behavior of the zooplankton. The laboratory feeding experiments with *Melaniris* were repeated, using a mixture of these "super-eyed" morphs and the normal large-eyed forms as prey animals. The result was that *Melaniris* reversed its behavior and now favored the small-eyed morphs with the super eyes (fig. 10). The laboratory feeding experiments showed that merely by changing the amount of black pigmentation in the area of the compound eye one could control the selectivity of the predator so that it fed on the most visible of the two *Ceriodaphnia* morphs. Further experiments showed that this manipulation had a similar effect on other prey species as well (Zaret 1972b).

As another example, Kislalioglu and Gibson (1976) showed that the 15-spined stickleback, *Spinachia spinachia* (Gasterosteidae), selected

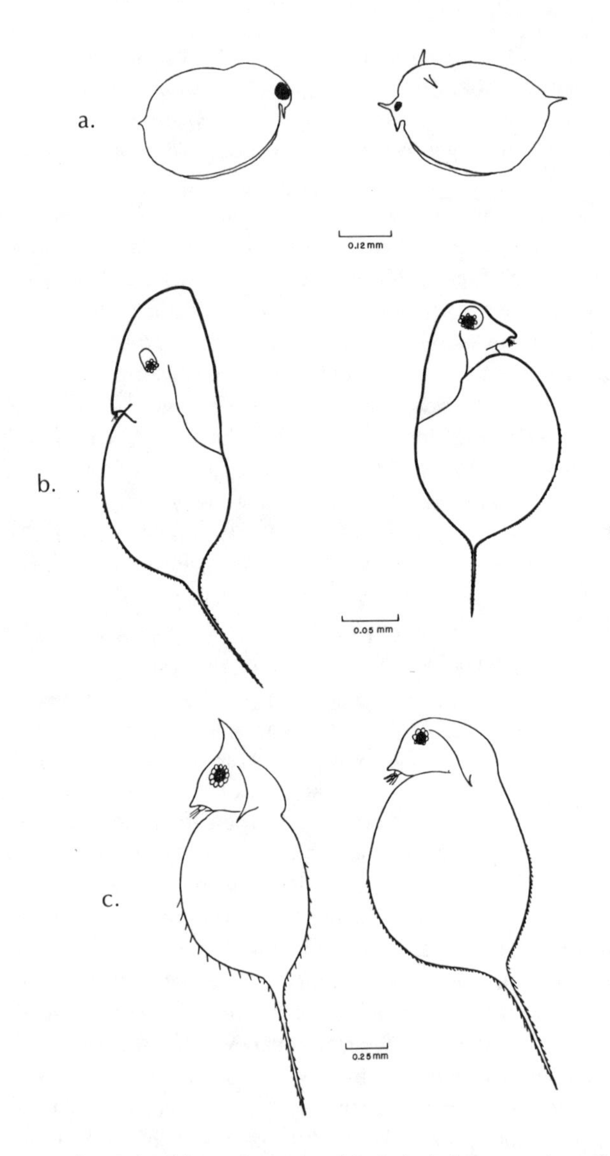

Fig. 9 a. Unhorned and horned morphs of *Ceriodaphnia cornuta* from Gatun Lake, Panama. (From Zaret 1972a) b. Helmeted and unhelmeted morphs of *Daphnia galeata mendotae*. (After Brooks 1957) c. Helmeted and unhelmeted morphs of *Daphnia lumholtzi* from Lake Albert, Africa. (From Green 1967)

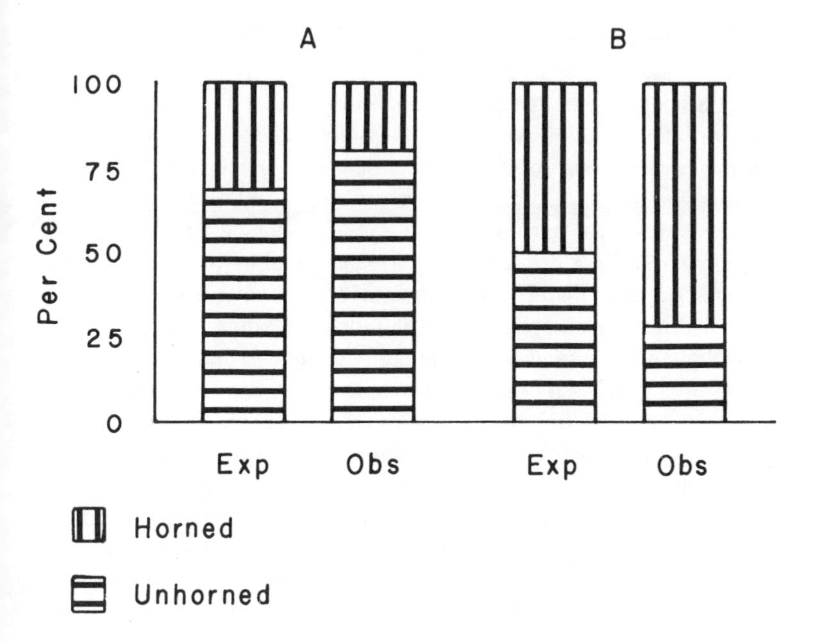

Fig. 10. Expected and observed predation frequencies in feeding experiments of *Melaniris chagresi* on a mixture of two *Ceriodaphnia cornuta* morphs: *A*, unhorned versus normal horned ($p < 0.02$, X^2); *B*, unhorned versus "super-eyed" horned ($p < 0.05$). (From Zaret 1972b)

darker individuals of the crustacean *Neomysis integer* (dipped in powdered carbon) over lighter (daylight-adapted) animals of the same body-size.

When one considers that fish vision depends upon contrast between the object and its background, the explanation for this result becomes obvious. A cladoceran having a black eye or other pigmentation will be seen more easily by a planktivore than one without a similar black area, and the largest individuals of the more visible species will be selected because they will have larger eyes. This helps to explain the results from the laboratory feeding experiments by Brooks (1968) and Werner and Hall (1974). Thus, this "visibility hypothesis," namely, that the prey item is selected by the predator according to total visibility, was proposed as an alternative to strict body-size selection by planktivores. An intensive test of body-size versus visibility selection on another cosmopolitan cladoceran, *Bosmina longirostris*, also supported the hypothesis that fish were selecting prey according to total visibility, including pigmentation, rather than body-size alone (Zaret and Kerfoot 1975).

It is not solely the compound eye, however, that is responsible for the total visibility of zooplankton prey. To paraphrase Brooks (1965), it is not the total size of zooplankton but the visible parts of the body that are the decisive factor. All aspects of its morphology can contribute. Mellors (1975) showed that female *Daphnia pulex* containing the dark, heavily pigmented ephippia are selected much more rapidly by the planktivore *Lepomis gibbosus* than those *Daphnia* of equal size that do not contain these dark spots (and see Nilsson 1960 for similar observations). Vinyard and O'Brien (1975) showed in laboratory experiments that bluegills (*L. macrochirus*) fed more on *Daphnia* having pigmented guts or high concentrations of red hemoglobins than on the clearer forms. Thus, areas of pigmentation are a liability to prey, which may explain why many lake daphnid species do not possess hemoglobin development even though this is correlated with higher filtering rates (Kring and O'Brien 1976), whereas daphnids inhabiting ponds that are normally without fish do (Fox 1948).

Stenson (in press) has found that the distribution of the aquatic larvae of two species of the phantom midge, *Chaoborus* (see later), is correlated with visual predation by fishes. One species, *C. flavicans*, is encountered in Swedish lakes with fish present, whereas *C. obscuripes* appears only after the fish have been removed. The species found with fishes has a significantly smaller eye; the other species, in addition to a large eye, has a dark head region, resulting from heavier pigmentation of the mandibles, and a darker yellowish-brown body. Laboratory feeding experiments by Stenson with *Chaoborus* and fish predators showed a significant preference by fishes for the darker appearing midge larvae.

All aspects of the prey's morphology may be expected to contribute to its visibility, including eye pigmentation, body pigmentation, gut pigmentation from ingested algae, and the chitinous body carapace as well. In the genus *Daphnia*, where total body-length can be as much as ten times longer than in the genus *Bosmina*, the influence of the carapace is accordingly more important in determining total visibility and the consequent predilections of the planktivores. Even in large *Daphnia*, however, those individuals with a reduced compound eye experience a lower rate of predation by planktivores. For instance, many *Daphnia* species undergo a seasonal pattern termed "cyclomorphosis" (see figs. 9a, 9b, and 9c), in which various aspects of their morphology change with succeeding generations within clones (see Brooks 1946; Hutchinson 1967 for a review).

Field and laboratory experiments have shown that morphs produced when fish predation is most intense experience a lowered rate of predation from fishes relative to the preceding morphs (Brooks 1965; Green 1967; Jacobs 1967). In all cases examined so far, the forms with the lowered rate of fish predation have always had a greatly reduced compound eye (Zaret 1972*b*). In fact, Green (1971) found that his previous work on helmeted and unhelmeted morphs of *Daphnia lumholtzi* in Lake Albert, Africa, could be reinterpreted in this light because those morphs that experienced lower predation by the lake planktivore *Alestes baremose* (Characidae) also possessed a smaller eye.

As another example, in *Daphnia longispina* individual eye size increases with pond altitude (Wawrik 1966), an inverse correlation with fish abundance.

Recent work has indicated possible parallels with marine systems in terms of the importance of prey visibility in the plankton. Merret and Roe (1974) found that the copepods *Pleuromamma piseki* and *P. gracilis* have a pigment spot on the side of the metasome that makes them very conspicuous, which may help explain why they are the most important food source of the gonostomid fish *Valenciennelus tripunctulatus*. Arthur (1976) concluded that the abundance of the copepod *Microsetella* in jack mackeral (*Trachurus symmetricus*, Scombridae) was explained by the conspicuous colors of this crustacean.

When the concept of body-size selection by planktivores was first developed, the *effect* of planktivore predation, namely, the removal of larger species and their resulting replacement by smaller forms, was being confused with the actual *mechanism* of planktivore selection. In feeding experiments involving a planktivore and a single prey species, such as a cladoceran (Werner and Hall 1974) or a copepod (Brooks 1968), predation is directly proportional to body-size, the larger forms being taken preferentially. Because body-size and eye-size are usually significantly correlated in crustacean populations, the relationship between predation and body-size resembles the relationship between predation and eye-size. It is difficult to separate the two effects. It was shown in *Bosmina longirostris*, however (Zaret and Kerfoot 1975), that the strength of the relationship between eye-pigmentation diameter and body-length declines after predation. This is due to the increased proportion of large-bodied but small-eyed prey remaining in the population (fig. 11). If predation occurred strictly according to body-size, the largest individuals of the populations would be removed, resulting in a truncated distribution, but the significant correlation between

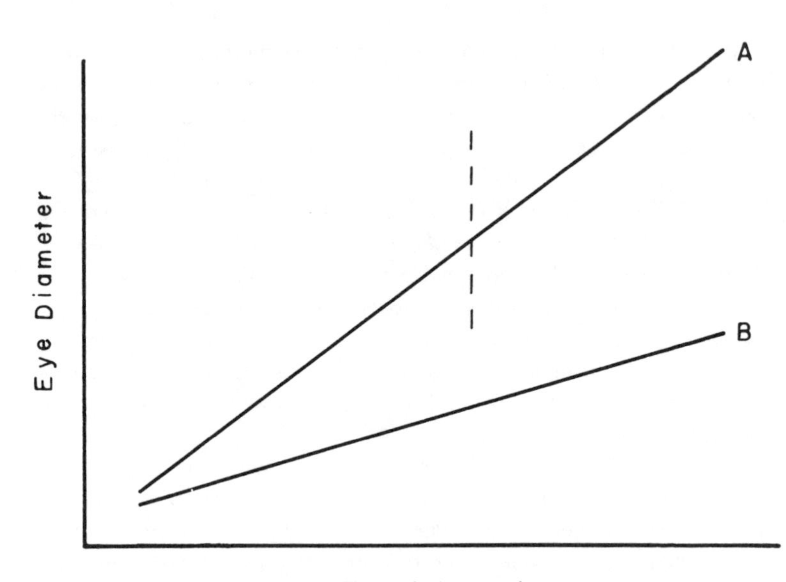

Fig. 11. Comparison of two possible regression lines of prey eye diameter to total body-length: *A*, normal regression slope remains if predator selects according to prey body-length but results in truncated prey distribution (signified by dotted line); *B*, change in regression slope if predator selects according to eye diameter, because only small-eyed prey remain.

eye-pigmentation diameter and body-length for the remaining individuals would remain strong. If the predator were selecting according to eye-size, however, it would remove many large-sized prey items because of the correlation between eye-size and body-size. It would also remove small, large-eyed animals, resulting in a larger number of large-bodied, small-eyed animals than before, and a diminished strength of the previous correlation. Through this correlation evaluation one can distinguish between these two modes of predator selection and test the importance of total prey visibility to predator electivity.

Thus, when planktivores feed on zooplankton, removing the most visible, largest individuals from the population, the result is a lowered mean body-size for the prey population. When we examine the stomach contents of the fishes we will find also that they contain animals of a much greater mean size than that of the remaining natural prey population. This finding, however, is due to planktivore selection based on *total* visibility—

of which body-size is only a single consideration, although an important one. As long as eye-size and body-size are significantly correlated, even if fish selected *only* according to eye-size, the result would be a reduction in mean body-size of the prey population.

As further support, one has only to consider a natural zooplankton assemblage, including cladocerans and copepods, and observe that it is the smaller cladocerans that are removed first by fishes in preference to the larger copepods. For example, in Brooks's 1968 feeding experiments with *Alosa*, the zooplankton collection used in the feeding experiment included a large cladoceran (*Daphnia catawba*) of mean mature instar body-length 1.2 mm and an even larger calanoid copepod (*Epischura nordenskioldi*) of mean length 1.68 mm. In the initial several minutes of feeding by *Alosa*, all of the cladocerans were eaten first, followed by the larger copepods. This preference for the smaller cladoceran over the larger copepod has been shown also by Ivlev (1961), as well as suggested in other studies (Berg and Grimaldi 1966; Keast 1970).

Oftentimes, smaller cladocerans are eaten regularly even though there are many larger cladocerans present in the areas where the fish are feeding (Burbidge 1974). By way of example, figure 12 presents the common limnetic crustacean zooplankton of Gatun Lake, Panama, drawn to scale. It is clear that the largest species are the copepods *Diaptomus gatunensis* and *Eucyclops agilis*, whose adults reach a mean size of approximately 1.1–1.3 mm. Following this in size are the cladocerans *Diaphanosoma brachyrum* (1.1 mm), *Moina minutus* (0.6 mm), *Chydorus eurynotus* (0.55 mm), *Ceriodaphnia cornuta* (0.35 mm), and finally the Bosminidae, including *Bosmina longirostris* (0.3 mm), which is the smallest. If the local planktivore, the atherinid *Melaniris chagresi*, were feeding according to body-size, one would expect it to prefer as prey the species in the order listed. In fact, the opposite is true. *Melaniris* exhibits the highest electivity for the smallest species, *Bosmina*, followed by *Ceriodaphnia*. The other two cladocerans, *Diaphanosoma* and *Moina*, are rarely eaten, nor is the largest crustacean, the calanoid copepod *Diaptomus*. The relative visibility of these four species—*Bosmina* and *Ceriodaphnia* have relatively large eyes in relation to their bodies, whereas both *Moina* and *Diaphanosoma* have relatively small eyes—explains this predator electivity.

Other workers also have found that body-size considerations do not adequately explain the selection of prey by planktivorous fish. In Hutchinson's 1971 study of the alewife, *Alosa pseudoharengus*, in small lakes in the Adirondack Mountains of New York, it was reported that *Bosmina* was

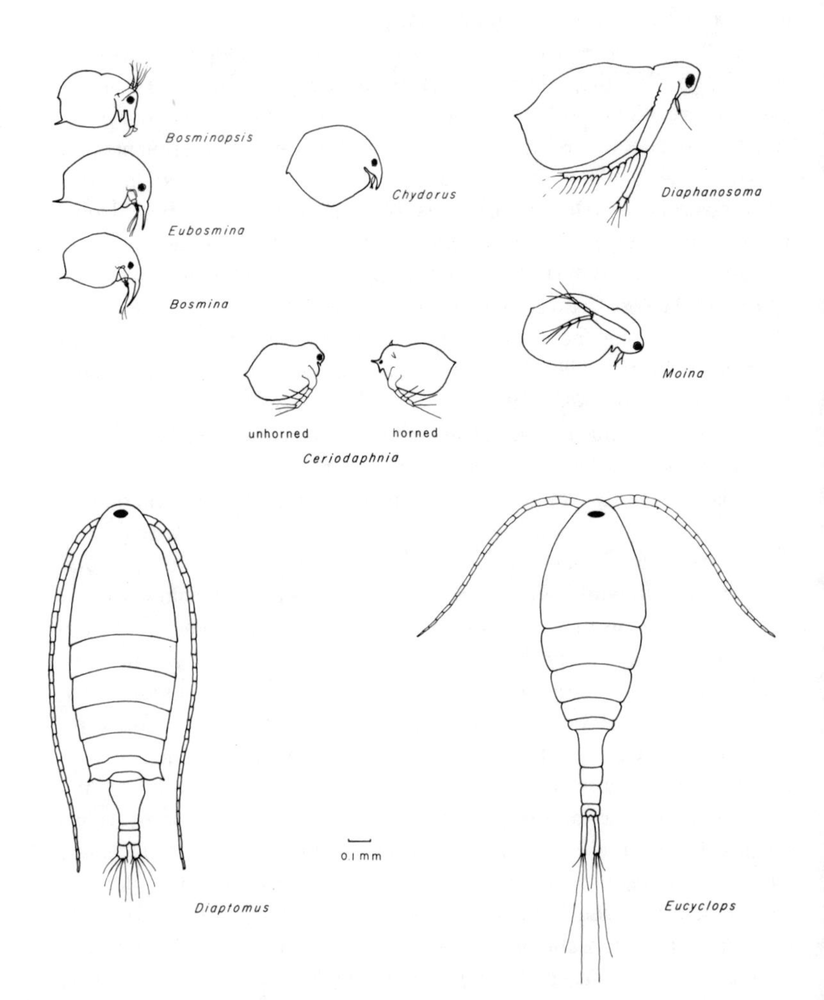

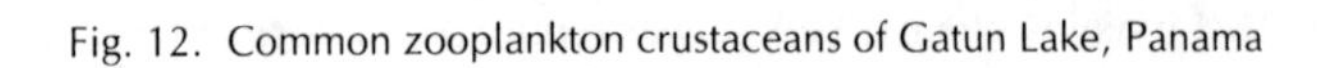
Fig. 12. Common zooplankton crustaceans of Gatun Lake, Panama

the prey most commonly taken even though larger species were available, including *Holopedium* and *Cyclops*. The average body-size of *Bosmina* during the five months of the study, May through September, was 0.31 mm; that of *Holopedium*, 0.61 mm; and for *Cyclops*, 0.53 mm. In this case, *Holopedium* is probably less preferred because of its outer gelatinous matrix, although it has been found to be a significant prey item for fish in other lakes (Stenson 1974; O'Brien 1975). *Cyclops* may avoid capture in part because of its ability to dart rapidly, which is probably most effective when other prey are available.

A study by Burbidge (1974) indicated that *Alosa aestivalis*, the blueback herring, strongly preferred *Bosmina longirostris* over *Diaphanosoma brachyurum* even though the latter was more than twice the total length of the smaller cladoceran. *D. brachyurum* is neither distasteful nor does it possess a rapid escape ability; rather, it is a very transparent species with a greatly reduced eye-pigmentation diameter. In addition, *Diaphanosoma* possesses a motion component (see following section) that may reduce its conspicuousness. The importance of total prey visibility as a key to planktivore electivity is stressed to avoid repeating past confusions between mechanism and effect in developing a comprehensive planktivore model.

Motion

Although the role of visibility in predator selection has been documented, a second important determinant of predator selection, namely, prey motion, has been much more difficult to test and is still not well defined. The ability of fishes to discern individual objects from among many moving in a visual field is most developed in planktivores (Protasov 1968), although the presence of nonmoving food objects, such as copepod eggs, in fish stomachs indicates that movement is not always essential for fish feeding (May 1970). Some fish species, such as the common goldfish, *Carassius auratus* (Cyprinidae), not only respond to motion per se, but actually possess receptors that distinguish fast from slow speeds and that the fish uses in different behavioral responses (Ingle 1968). The ability of the fish to perceive the prey, however, still depends on the contrast between the object and its background (Hemmings 1966; Hester 1968), which is proportioned to ambient light intensity. The visual pigments of the fish's eyes are adapted to maximize this contrast according to the light regime in its natural environment (Lythgoe 1966; Munz and McFarland 1973). Thus, if an object is visible to the fish at 10.0 m but not at 10.1 m, no amount of motion will make that object visible to the fish at 10.1 m. Once the object is within the visible range, however, motion may increase its conspicuousness and thus affect the probability that the prey will be noticed by the fish. It seems reasonable to assume that once the prey is within their visible range, predatory fish respond differently to different motions, probably mediated by the direct stimuli of the surface on the retina (see references in Eggers 1977).

Several field and laboratory studies have documented the effect of prey motion on predator electivity. Lindstrom (1955) recognized the importance of motion when he observed in the laboratory that the fry of char, *Salvelinus alpinus*, eat pike fry when the pike move but lose interest when pike are motionless. Boulet (1958, cited in Curio 1976) has suggested that fish recognize *Daphnia* by their mode of locomotion; and Braum (1963) implied the same for *Coregonus* feeding on copepods. As another example, in his study of the three-spined stickleback, *Gasterosteus aculeatus* (Gasterosteidae), living in streams along the coast of Washington State, McPhail (1969) found that motion was one of the key attractants for its natural predator, the endemic western mudminnow, *Novumbra hubbsi*. Kislalioglu and Gibson (1976) showed that the fifteen-spined stickleback, *Spinachia spinachia*, selected moving prey in preference to stationary prey (the crustaceans *Neomysis integer* and *Prawnus flexuosus*) in almost all cases. Finally, Ware (1973) showed that rainbow trout, *Salmo gairdneri*, were able to locate moving prey more successfully than stationary prey with otherwise identical properties.

To examine the importance of motion, Ware designed laboratory feeding experiments to measure the reactive distance of rainbow trout to inanimate "prey" (commercial chicken liver pieces). He placed the prey on a small platform in the aquarium with the fish; an exterior motor moved the platform, which was 1 cm from the tank floor, through an arc at the rate of 2 cm per second. For any of the prey sizes used, the trout detected the moving prey sooner than when it was stationary (fig. 13). These studies, and there are many others as well, support the conclusion that prey species in motion are most vulnerable to fish predation and suggest that survival tends to favor those prey species that move the least.

Prey conspicuousness may possibly enable fish to form what Tinbergen (1960) has called a "searching image."* Because of ambiguities, the term "searching image" has been modified subsequently to refer specifi-

* According to Curio (1976), the idea of a "searching image" was conceived first by J. Uexkull and G. Kriszat (1934) and revived by L. Tinbergen (1960).

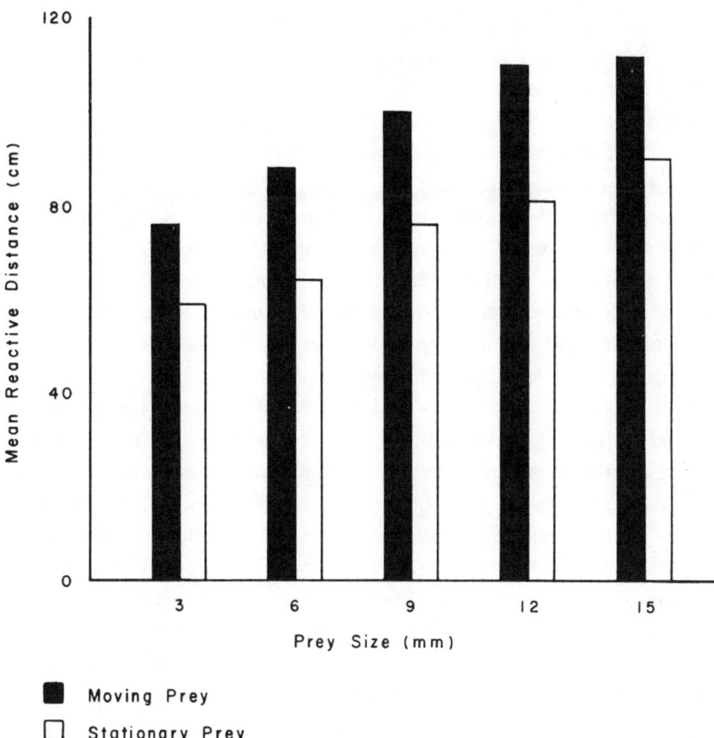

Fig. 13. Effect of prey motion on predator mean reactive distance. *Salmo gairdneri* on artificial food. (From Ware 1973)

cally to changes in feeding behavior related to vision or visual clues (Dawkins 1971). (For a review and discussion of the relative merits of the concept, see Krebs 1973.) It is clear that the phenomenon itself is real; the extent to which a searching image is employed by various predators is still being debated (Krebs 1973; Paulson 1973).

Planktivores may facilitate prey-capture success by learning to recognize prey items according to their motion. That is, according to accumulating empirical evidence from experiments, experience can influence vertebrate prey-capture success (e.g., de Ruiter 1952; Holling 1959; Beukema 1968; Croze 1970), a concept that has also been suggested for fish (Popham 1941, 1942). The searching-image concept is supported by Rosenthal and Hempel (1970) in their study of the feeding runs by herring; when surrounded by a diverse assemblage of planktonic prey, the herring feed exclusively on one type of prey and follow with an exclusive run on another. Other studies of the searching-image concept include Beukema (1968) with *Gasterosteus* and Ware (1971) for *S. gairdneri* (and see Nilsson 1978). Whether these studies have demonstrated the searching-image concept in the strictest sense has been questioned (e.g., see review and discussion in Krebs 1973). Fish do nevertheless depend on prey motion for feeding and experience does play an important role. As Ware (1972) concludes, this type of response might act as a positive feedback to improve the efficiency of vertebrate predators, whether we can call this a searching image or not, and it may help to explain the mechanistic basis for fish taking certain prey species over others.

This last point and the fact that fishes differentiate among available prey species on the basis of motion are probably critical to an understanding of planktivore feeding. That is, we may envision a planktivore swimming until it locates a cloud of zooplankton prey, and among the many particles of that cloud, the species with the most conspicuous motion component will be the focus of the fish's interest. It is most likely that larger prey will be more conspicuous in their movements, but this is not always the case, as previously mentioned for *Bosmina* and *Diaphanosoma*. Prey of similar size may be distinguished from each other by specific behavioral actions. Among all the individuals of this preferred-motion prey population, some will be more visible than others (the largest, most heavily pigmented) and will be consumed first. Once this species is removed from the prey cloud, other forms may become the center of the planktivore's interest. This interpretation, incorporating an interaction between motion and visibility, will have to be tested more thoroughly. However, body-size and total visibility are still not sufficient in themselves to explain certain sets of observations (Zaret in press).

Models

One of the first attempts at developing a quantitative model to describe the selection of planktivores for prey was Werner and Hall's (1974). They examined the hypothesis that fish choose prey to maximize their caloric intake per energy expenditure in prey capture (E/t; energy per unit time). This foraging study offered the generalist *Lepomis macrochirus* different-sized instars of the cladoceran *Daphnia magna* in artificial pools. Werner and Hall empirically determined functions for predator search time, handling time, and energy return for each prey instar. Based on these parameters, the bluegill fed in a pattern of foraging that seemed to optimize energy gain. At low prey densities, food items were taken and consumed as encountered. With increasing densities of prey, however, the predator concentrated on the largest prey sizes out of proportion to abundance; and whenever large sizes were abundant, smaller prey sizes were omitted from the diet. A decrease in the breadth of the fish's diet occurred, therefore, with higher prey densities, which corresponds to Ivlev's (1961) findings that fish selectivity increases with increasing prey density. Werner and Hall related this change in diet to cost of prey capture in terms of time and energy in a pattern termed "optimal foraging." The mechanistic basis for the fish choice was not considered.

A study of planktivore feeding (Confer and Blades 1975) based on laboratory experiments with *Lepomis gibbosus* attempted to provide the mechanism needed to explain planktivore preferences for larger prey sizes. Confer and Blades developed the concept of the fish's field of search, equating it to a cylinder with a radius equal to the distance at which the fish first saw a given prey item (first termed "reactive distance," RD, by Ware 1971). The authors argued that a predator moving through a cylinder of water searching for prey is more likely to encounter larger prey because their greater surface area will be detected more rapidly than smaller prey. Once encountered, the prey will be pursued and possibly consumed, depending on the prey's ability to escape. This prey encounter frequency model is based on the reactive distance of a given prey, which is largely a function of prey size, and on the ability of the predator to capture different types of prey. When the probability of encountering two prey at the same time is small, prey should be taken in proportion to the square of fish reactive distance times the density of a given prey. This model appears to be quite accurate at relatively low prey densities.

A more quantitative model based on similar assumptions was developed recently by Eggers (1977). Again, the predator first locates prey according to body-size and the predator's field of search. Prey encounter is the product of the predator's visual field and prey density (Holling 1966). This formulation can be used to predict specific distributions of prey consumption by predators and can be very useful in analyzing resultant prey distributions.

All three approaches fit data exceedingly well when fish are feeding on instars of a single prey species. Of course, when fish are selecting from different-sized instars, similar in all respects except body-size, it would be surprising to find fish distinguishing on any other basis. Major discrepancies with these models occur, however, when mixed assemblages of prey species are involved. Under these circumstances the models diverge from empirically determined field data. Some smaller-sized prey groups, such as cladocerans, are preferred by fishes over larger-sized prey groups, such as copepods (Ivlev 1961; Brooks 1968; Berg and Grimaldi 1966); and within prey groups, again such as cladocerans, often smaller species are taken over much larger ones (Hutchinson 1971; Burbidge 1974). The selectivity of planktivores (for *Coregonus*) was perhaps first considered by Wagler (1941), who suggested that selection was determined by various possibilities: taste, capturability, relative abundances, and conspicuousness resulting from prey size.

To account for the preferences of cladocerans over copepods one might argue that copepods can escape the visual field of fishes better than cladocerans and are thus less likely to be detected (Confer and Blades 1975; Drenner, Strickler, and O'Brien 1978), or that cladocerans, being round, are more easily seen by fishes than are copepods of similar length (Eggers 1977). A rebuttal is that even though copepods may be more difficult to capture, as measured by the success of fish attacks, planktivores such as *L. gibbosus* can learn to capture copepods with 80 percent efficiency within a relatively short period of time (Confer and Blades 1975). Second, although copepods may have an elongate shape, their antennae are normally at right angles, giving them a much broader surface area, which may be important to fish attraction. Finally, it has been demonstrated that small copepods elicit a shorter reactive distance than cladocerans of the same size (Confer and Blades 1975). This cannot be explained by consideration of body-size alone.

To improve upon the previous formulations of planktivore models and remedy these two major weaknesses, O'Brien, Slade, and Vinyard (1976) proposed a mechanism for fish selection. They hypothesized that at the instant the fish initiates its search for food, it selects the largest appearing prey, based on a combination of absolute size and proximity to the fish. This work also examines one of the major shortcomings of Werner and Hall's 1974 study in that it asks how a fish can distinguish between a large but distant prey and a small but nearby one?

The model of O'Brien, Slade, and Vinyard is based on laboratory experiments with bluegills in wading pools, following the methods of Werner and Hall. The authors present prey of different sizes at different distances from the predator, calculating prey height and dividing by the distance from the fish to obtain the arc tangent. Using this value to determine which prey appears largest, their data indicate that the fish do select the largest appearing prey regardless of actual prey size. Because large prey are more likely to appear in the fish's visual field, there is a greater chance that these forms will appear larger and will be eaten first. Their model predicts the data of Werner and Hall (1974) but is based on very different assumptions. One advantage of this model is that the problem of different prey visibilities can be obviated by saying that pigmentation merely makes the prey appear larger (by means of its being seen more easily or, using Ware's concept, reducing the reactive distance for this item). Thus, instead of having to deal directly with two distinct components of prey attractability, namely, visibility and motion, each of these two categories can be valued according to how each would affect what the authors call apparent size.

Certain observations remain that cannot be incorporated easily into the O'Brien, Slade, and Vinyard model. In their model, fish feeding falls on the largest forms, reducing their numbers, until the large forms become so rare that smaller prey are as likely to enter the predator's visual field and be selected. This is a gradual change, however, and one would expect that the largest sizes would never reach extinction but become rarer, the fish gradually switching more frequently to the next smaller sizes. We do know, however, that fishes cause the extinction of daphnid species precisely by removing all the individuals of given size classes (Galbraith 1967). In addition, the model does not explain how fish, given the choice of two different-sized prey species, can select one form to the exclusion of the second, even when the two forms are of similar apparent body-size. This has been shown by Green (1967) for the planktivore *Alestes baremose* feeding on the two morphs of *Daphnia lumholtzii*, where the fish selected 100 percent of the unhelmeted form. This degree of selection cannot be explained by the apparent-size model because at least some of the time other prey should be expected in the stomach of the predator. The same argument can be applied to Galbraith's study, where prey under 1.3 mm were never taken although very abundant. One would have expected at least some of these small forms to be taken, unless the predator is actively avoiding them or using some other mechanism for prey location.

The limits of mathematical flexibility seem to have dictated, in a sense, the approach used for many planktivore models and, consequently, models based strictly on body-size are especially attractive because they can be developed in a quantitative manner. This is especially evident in the model of O'Brien, Slade, and Vinyard, which basically takes defined areas of known prey interaction with predators, such as visibility and motion, and translates them into one single response, apparent body-size, in order to extend the model originally presented by Werner and Hall. One can bend the data only so much until the flexibility of the model limits the advancement of knowledge. Many more investigations of the interaction of body-size, visibility, and prey motion are necessary before a model can be constructed that will explain the wealth of data available from field studies.

3. More on Planktivores

PLANKTIVORE FEEDING EFFICIENCY

A fish's efficiency* at feeding on zooplankton can be predicted from its morphology. A comparison of the morphology of five common freshwater planktivores—(a) an atherinid (silverside), (b) a clupeid (alewife), (c) a coregonid (whitefish), (d) a salmonid (rainbow trout), (e) and a centrarchid (bluegill) (fig. 14)—will best illustrate differences in feeding efficiency that result from various morphologies.

Going from the most efficient to the less efficient planktivores, a change occurs in the eye and in its position relative to the head, in the position and opening of the mouth, in the placement of the fins, and in the general body form. In atherinids, the eye is placed high on the head and forward in a position that allows the fish to look up at the prey silhouetted against the natural sunlight, thus providing maximum contrast, the key to fish vision. The lower jaw (mandible) is extended somewhat forward so that it takes but a quick nip, the opening of the buccal area creating a slight vacuum, to suck in the prey item. The pectoral fins, placed high on the body, and the two sets of dorsal fins provide the great mobility and control that allow the fish to track the prey's darting moves. Finally, the body itself is shallow and delicate, so that the fish may comfortably feed just below the water surface. In addition, the silver color pattern of atherinids makes them less conspicuous to predators in open-water situations (Denton 1971). After considering this most efficient planktivore, note the other ex-

* Efficient planktivores are those that can exist on small prey items, in the range of 0.3 mm, when other species are forced to switch to alternative food sources such as insects.

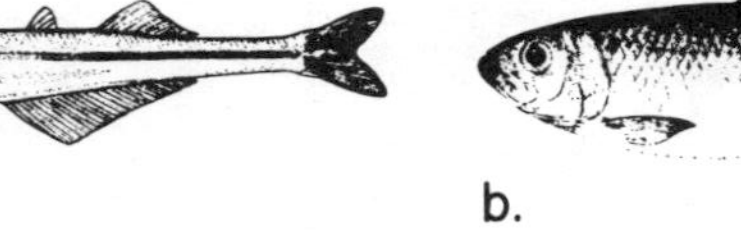

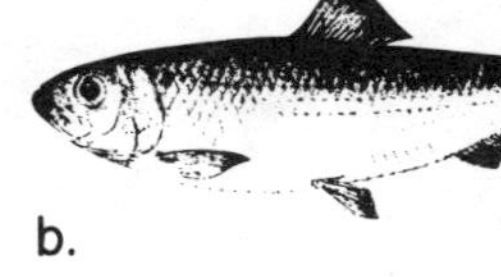

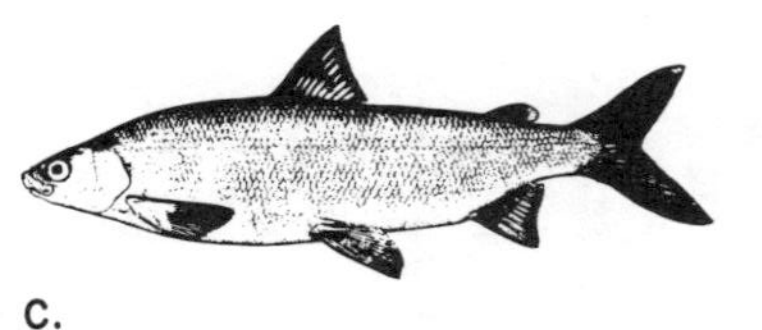

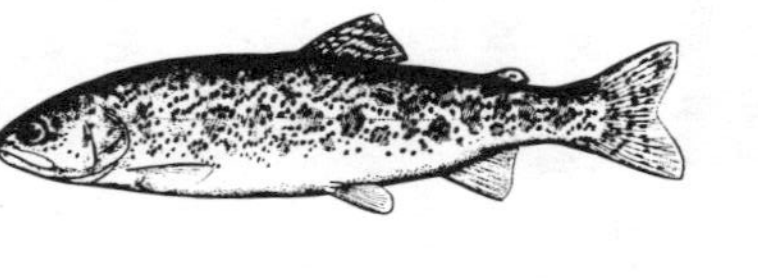

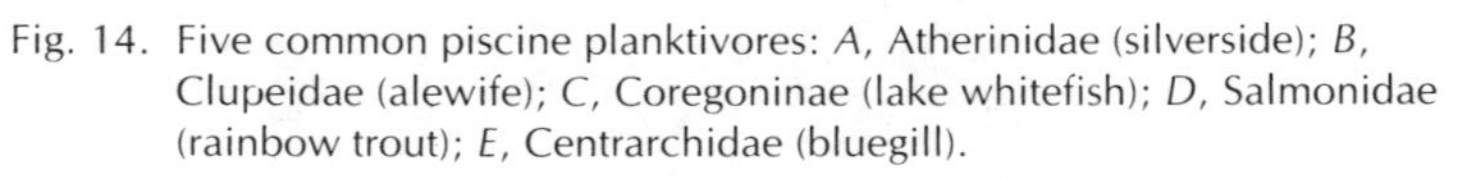

Fig. 14. Five common piscine planktivores: *A*, Atherinidae (silverside); *B*, Clupeidae (alewife); *C*, Coregoninae (lake whitefish); *D*, Salmonidae (rainbow trout); *E*, Centrarchidae (bluegill).

treme in the centrarchid, the bluegill, only the young of which are strict planktivores. The eye is located downward, in a position to see items directly in front of it rather than above, a compromise that allows the bluegill to feed on surface or benthic animals. The mouth is fairly even-jawed and strong, an evolutionary response that enables the fish to hold and feed on large food items. The gape is large, providing a great area of negative pressure upon opening, for swallowing large items whole; the fins are lowered and moved posteriorly; and the entire body is deep, composed of the musculature needed for rapid movements to allow feeding on fast-moving prey, including other fish. As figure 14 illustrates, such changes in morphological structures are adaptive for the particular environment of the fish.

The most successful (dominant) planktivores found in fresh waters—the Salmonidae, Coregonidae, Clupeidae, and Atherinidae—have a familial background derived from marine systems. One notable exception is the Cyprinidae (e.g., carp). The fact that many of the most efficient planktivores have marine affinities in their recent past may have several reasonable explanations. First, marine fishes have had a much longer time to evolve a plankton-feeding ability compared with freshwater relatives, as all but a few of the world's present lakes are relatively young (i.e., less than 1 million years old). Second, the more constant seasonal supply of zooplankton in marine systems allows these fishes to specialize on them throughout the year (Cushing), whereas in most freshwater systems zooplankton abundances drop during certain periods of the year and planktivores are forced to switch to other food items (Keast 1970). This is probably related to nutrient cycle differences between these two aquatic systems. Finally, as fish increase in size in marine systems, larger zooplankton and even schools of fishes may serve as prey food, whereas most limnetic zooplankton are restricted to sizes of less than 5 mm, with only a few of the largest predatory invertebrates attaining sizes of 10 to 20 mm. Therefore, as they grow, most planktivores are forced to switch to larger food items, most of which are not plankton.

As a result of these differences in the basic organization of marine versus freshwater systems, the commonest freshwater fishes of North America (e.g., centrarchids, including crappies, bluegills, sunfish, and bass; ictalurids, including various catfishes; cyprinids, including shiners, chubs, and dace) are generalists. They are capable of feeding on plankton, especially as juveniles, but are also capable of switching to insects or other prey as these become available in lakes, unlike many marine planktivores that feed on zooplankton (or phytoplankton) exclusively throughout their lives. When considering planktivores, this should be remembered whether we are dealing with specialized fish, most of which have past marine affinities and are highly efficient, or with facultative fish, including many just mentioned that are generalists, adapted strictly to the fluctuations of food abundance in lakes (see Confer et al. 1978). It is not clear how well generalizations from studies of the bluegill, *L. macrochirus*, a plankton generalist, will hold for other planktivores.

A PLANKTIVORE MODEL

A schematic model depicting how a piscine planktivore functions in obtaining its prey may be useful to suggest some of the interesting factors that help to determine "predator catch." Predator catch designates which species and how many of each will end up in the planktivore's stomach after a given feeding period.

Figure 15 is a single flow-chart model based upon information from the various studies discussed earlier. This model leaves out certain details and makes some questionable assumptions; I will refer to the sources on which the assumptions are based or will discuss them. The effect of other predators on the planktivores and the effect of time have been omitted for purposes of simplification. Obviously, the amount of time a predator spends feeding will affect such aspects as prey density, which itself will have other consequences on total Predator Catch. For the purposes of this model, however, I will assume a very short time interval, so that the effect of time is negligible. This simplifies the explanation considerably, although it should be remembered that feeding plays a considerable role in the life of any predator.

This model relates to three areas—the characteristics of the physical environment, the characteristics of the predator, and the characteristics of the prey—that together determine Predator Catch.

Physical Environment Characteristics

The effect of the sun in the rectangle entitled "light condition" includes the intensity of solar radiation as determined by latitude, the inclination of the earth and ray angle (measured by time of day), and also wavelength as affected by such natural filters as clouds or particulate matter in the atmosphere. Light condition plus water transparency, determined by biotic and abiotic particles in the water column as well as water surface turbulence,

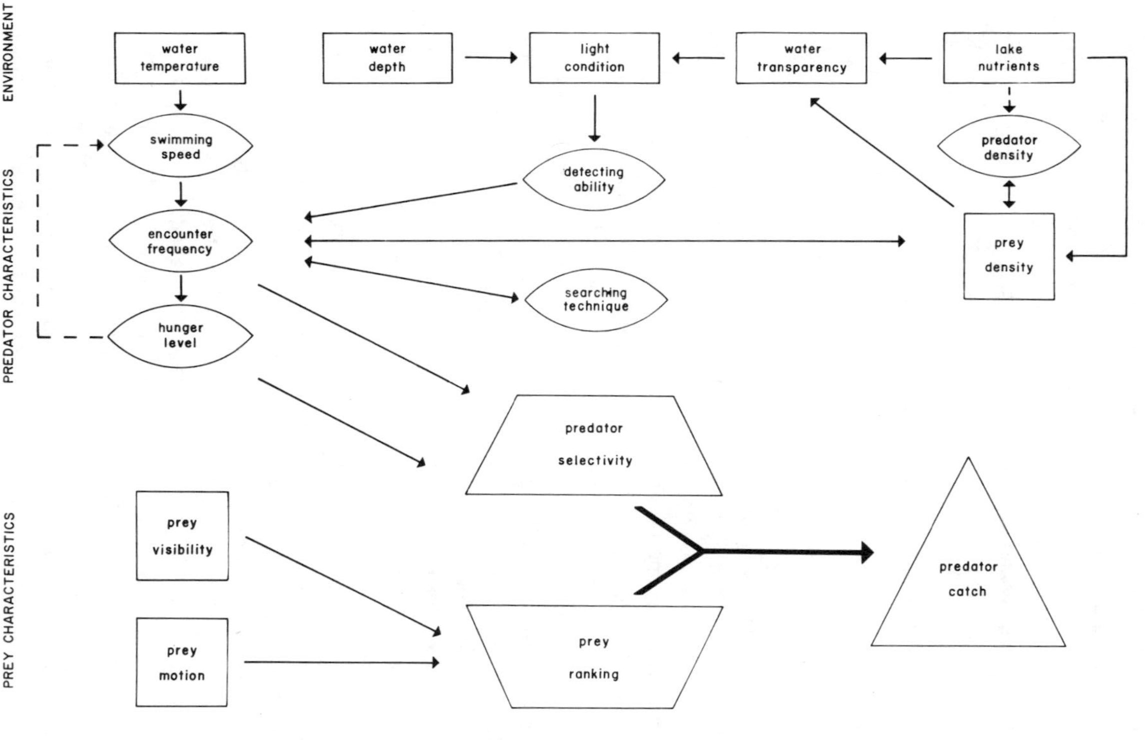

Fig. 15. Planktivore model (the bold line identifies the major determinants of predator catch; the regular solid lines signify proven relationships; the dotted line signifies a questionable relationship).

and water depth (i.e., water may be very clear but shallow, setting an upper limit) will determine ultimately the detection ability of the planktivore. Water transparency will be affected greatly by lake nutrients (lake trophic state), which in turn are regulated by allochthonous (from outside the lake basin) nutrient sources as well as autochthonous (within the lake) biological activity and nutrient regeneration. There is also a feedback loop in that lake nutrients will affect prey density and prey density will affect detection ability. Water temperature will affect swimming speed and, in turn, categories dependent on this predator characteristic.

Predator Characteristics

The planktivore characteristics important to this model include predator density, which, in turn is likely to be affected most by prey density, although the predator population could be limited by habitat availability, spawning sites, or other factors critical to its biology. It is conceivable that predator density is also limited by insufficient vitamins or the presence of trace elements in the lake nutrients category, although this has not been documented. Predator density can be limited also by a species' own predators, although this is ignored in the model for simplification.

Predator hunger level is also important. Ivlev (1961) and others (Kislalioglu 1976) have found it to be a significant determinant of predator selectivity, although Ware (1972) did not find it so in his experiments. "Selectivity" indicates the degree of discrimination by a planktivore, given an assortment of different prey items. Hunger level will be most affected by prey encounter frequency, which in turn relates strongly to prey density. Detecting ability will have a major effect on encounter frequency, the rate at which the predator encounters prey, as will planktivore swimming speed. The latter in turn may be regulated by planktivore hunger level (and water temperature): if a planktivore does not encounter prey, and its hunger increases, the planktivore may increase its swimming speed in search of food. Ivlev (1961) concluded that hunger did not affect the swimming speed of fish, but Holling (1966) questions the validity of Ivlev's conclusions based on his experimental methods and believes that the average speed of many predators is affected by hunger levels. Finally, planktivore searching technique relates to whether a planktivore optimizes energy per feeding encounter (see Schoener 1971, and a review of optimal foraging by Werner and Hall 1974), switches from one prey type to another (see Murdoch 1969, 1971; Murdoch, Avery, and Smyth 1975), and uses the searching-image technique (see Beukema 1968; Ware 1971). The particular searching technique used affects prey encounter frequency and prey density in that it depletes the resources in some predictable way (Orians, Charnov, and Hyatt 1976) and varies, depending on encounter frequency.

Of all the categories designated so far, those of encounter frequency and hunger level are the key determinants of selectivity (concluded by Ivlev 1961). Given the particular searching technique and a choice of prey types, how selective the planktivore will be is determined by encounter frequencies or hunger, which in turn are determined by the various interactions described. Whereas Ivlev (1961) used hunger and prey density as his determinants of selectivity, prey density is important only according to the density actually perceived by the predator, that is, the predator's encounter frequency, as opposed to what is measured by the experimenter.

Prey Characteristics

Visibility and motion are the prey characteristics that are important to planktivores. Prey visibility is determined by body pigmentation in the eyes, gut, eggs, or the body integument or carapace—that is, all aspects of the prey that decrease the reaction distance of the planktivore. Prey motion refers to all behavioral aspects that make the prey more conspicuous to the planktivore or that help the prey to escape once it is seen or detected by the planktivore. For planktivores relatively large in relation to the size of their prey, prey escape responses are of little value and the best strategy is not to move at all. Prey visibility and motion interact to determine prey ranking by the predator. That is, given a spectrum of prey possessing different combinations of visibility and motion characteristics, the planktivore will rank certain forms over others.

In the final result, it is the direct interaction of predator selectivity and prey ranking that results in predator catch. Once the prey ranking is established, the selectivity of the planktivore will determine whether it will take 100 individuals of prey item number 1, of caloric value 6; or 50 individuals of number 1 and 100 of prey item number 2, of caloric value 3; or some other combination of prey types with different caloric values (Covich 1972).

ELECTIVITY

The concept of electivity was championed by Ivlev (1961), whose book summarizes approximately ten years of research—mostly with non-"planktophages," as he refers to them—by Soviet scientists. Although today we may look askance upon the rigor of some of the methods used, in general most of Ivlev's results and conclusions are accurate and provide important stepping-stones in modern feeding ecology.

One of the basic questions about the feeding ecology of any animal is how to interpret the predator's stomach contents. Many formulations relating the material in the animal's stomach to what is "available" in the animal's environment have been devised. The real problem is, of course, the word "available." Many prey items are present, perhaps abundantly so, yet not available in a real sense to the fish because of the physical refugia available to the prey, other physical barriers, prey size, protective structures, distastefulness, protective coloration, camouflage, or the escape abilities of the prey.

Ivlev's expression for relating stomach contents to prey availability is:

$$E = \frac{r_1 - p_1}{r_1 + p_1}$$

where r is the "ration" eaten, or stomach contents of the fish, and p is the "proportion" of an item found in the environment or, in our case, in the plankton. Using this formula, the electivity (E) of a food item varies from $+1$ (eaten whenever encountered regardless of abundance in the plankton) through 0 (selection of the item on a strictly random basis, i.e., eaten in the same frequency as encountered in the plankton) to -1 (avoided by planktivore regardless of abundance in the plankton).

A second commonly used measurement of food selection is the forage ratio (FR), which is simply the ratio of the ration of food type to its proportion in the environment (Edmondson and Winberg 1971), or

$$FR = \frac{r}{p}.$$

This value varies from 0 (avoidance) through 1 (random feeding) to $\propto$ for the most positive selection. This measurement has been used when comparing electivities of more than one predator on the same food items from the same waters.

The two major weaknesses of these two commonly used measurements of predator electivity are, first, an abundant prey item might have an electivity of -1 if it can be seen or captured by the experimenter but not by the fish. In other words, we might consider this item abundant, but because of its ability to hide or disappear, it might not be at all "available" for the predator. This would give the item a negative electivity, even though it might be a highly favored or selected item if the predator had access to it. In addition, inclusion of this item will have a disproportional effect in electivity calculations. Thus, problems of prey accessibility or patchiness of prey distribution (e.g., plankton patchiness) could negate conclusions based solely on electivity measurements without other knowledge (see O'Brien and Vinyard, 1974, for a discussion of this problem).

A second caveat for electivity indices is their total dependence on abundances of the prey type in the plankton. For example, if we are comparing the predator electivity for prey species *A* versus species *B*, even though 90 percent of the predator's diet may consistently be food item *A*, the electivity for *A* may decrease over time simply because its abundance is increasing in the environment. The reader is encouraged to take Ivlev's formula and substitute different numbers for r and p to explore its limitations and uses. The type of problem just mentioned is discussed by Jacobs (1974), who proposes a variation of this formula to obviate these weaknesses in the electivity measurement.

Electivity is simply a quantitative measurement, relating what has been termed "predator catch" to what was in the predator's environment at the time. It has meaning only as a relative measurement; we use electivity to compare how the fish is eating species *A* in relation to species *B*. Although prey ranking, which is the relative desirability of prey types given equal numbers of each, is a fixed value, electivity for any given item changes with the density of the prey. This last aspect puts a premium on the biological knowledge and intuition of the experimenter.

In a recent paper Chesson (1978) has developed a measure, by normalizing Ivlev's original formula, which maintains all of the advantages but eliminates the second weakness. This new index will probably make all previous electivity indices obsolete.

NON-PISCINE PLANKTIVORES

AMPHIBIANS

Urodelans (Salamanders). Many bodies of standing water do not contain piscine planktivores. In these bodies the role of the gape-limited predator is often assumed by another important group of planktivores, salamanders. By far the best studied and most common salamander to assume this role is *Ambystoma tigrinum*, the tiger salamander. This species preys

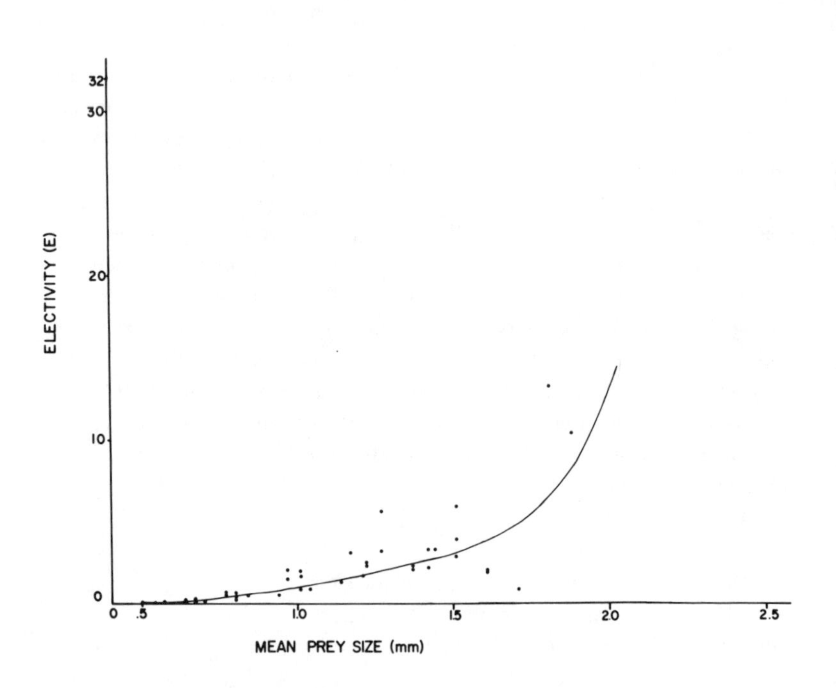

Fig. 16. Electivity curve of *Ambystoma tigrinum* from Colorado ponds, feeding on zooplankton prey. (From Dodson 1970)

on zooplankton, aquatic insects, and even on larvae of other sympatric *Ambystoma* species (Wilbur 1972). When feeding on zooplankton, *Ambystoma* shows the typical gape-limited electivity curve (fig. 16).

Salamanders do not normally persist in water bodies containing fish. As has been demonstrated, the introduction of trout can cause the local extermination of previously abundant aquatic *Ambystoma* populations (Burger 1950, in Sprules 1974). Under specific conditions, however, salamanders are able to coexist with predatory fish. Sprules (1974) found both salamanders and fishes in the same water body in several of the small ponds and lakes in the region of Vancouver, British Columbia. There the populations of *Ambystoma gracile* actually remained buried in the soft lake sediments during the daytime, evidently avoiding the fish, and emerged only at night to feed, a strategy comparable to vertical migration patterns in crustacean prey. In contrast to this unusual behavior, *Ambystoma* in other nearby ponds where fish were absent swam about actively and conspicuously during the daytime. Such lakes with the co-occurrence of fish and salamanders are unusual, and generally one finds an overall inverse correlation between the two, as Sprules (1974) demonstrated.

The best-documented case of the community effects of *Ambystoma* as a gape-limited predator comes from the alpine pond studies of Colorado, which have been undertaken by several workers (Dodson 1970; Dodson and Dodson 1971; Sprules 1972; Sexton and Bizer 1978). In the alpine regions above 900 m elevation, the ponds do not contain fish, probably because of a lack of access as well as a high frequency of freezing or desiccation of the ponds, most being shallow seepage ponds, many less than 1 m depth. In these transitory aquatic habitats, *Ambystoma tigrinum* exhibits a life history that enables it to take advantage of the food resources. This salamander maintains fully "metamorphic" populations. During spring or early summer, the adults lay eggs that hatch soon after to form aquatic larvae. Normally, these larvae persist in the pond for a single summer, after which they metamorphose into terrestrial adults. There are variations, however. Larvae may remain in the pond for more than one summer, overwintering and remaining as aquatic larvae up to as many as four summers before they metamorphose. At the other end of this continuum, some larvae never metamorphose but actually mature in the aquatic form and produce larvae within the lake. (In the famous Mexican subspecies *A. tigrinum mexicanum*, which exhibits permanent aquatic adults, the larvae are called axolotyls.) This aspect of the amphibian life history is referred to generally as "neoteny" and is commonly found in many salamander families. Strictly speaking, neoteny is the retention of juvenile characteristics past the normal period of maturity, whereas the previously described phenomenon of sexually mature neotenic individuals is correctly referred to as "pedogenesis" (see Dent in Etkin and Gilbert 1968). A further point of confusion is added in the Ambystomidae, because they actually possess facultative neoteny.* Even after having produced larvae for many years, pedogenic individuals can be induced to metamorphose by crowding, a reduction in food levels, or starvation (see Adolph 1931, D'Angelo et al.1941, Etkin and Gilbert 1968). The two cycles, metamorphic and pedogenic, are thus two ends of a continuum, with each capable of entering the other cycle given the proper environmental stimuli. Because the aquatic forms of *Ambystoma* depend on zooplankton and insects for their food, the response to starving or crowding can be seen as an

* See Gould, 1977, for a discussion of pedogenesis versus neoteny: chapt. 7; chapt. 8:299–302; and chapt. 9:319–24.

adaptation that allows them to leave the pond to find food some other way when available resources are or will be in decline.

In general, shallow ponds contain only metamorphic salamander populations because of the likelihood that the ponds will dry up or freeze during the year. Pedogenic individuals, at a disadvantage in shallow ponds, are found almost exclusively in deeper, permanent water bodies. Furthermore, cannibalism within this species in the deep ponds serves to inhibit the successful introduction of the small larvae from metamorphosed *Ambystoma*. In some cases, however, both forms may be found together, at least temporarily.

In the absence of piscine planktivores, salamanders such as *Ambystoma* can potentially fill this vacant niche and exploit the available aquatic resources in a manner nearly identical to fish. In fact, it has been suggested that aquatic newts of the genus *Triturus* may actually compete with fishes when together, as evidenced from the overlap of diet between the two groups (Avery 1968).

The characteristics that qualify *Ambystoma* as "a fish in amphibian's garments" are the following:

1. They exhibit a gape-limited electivity curve for zooplankton, with the upper limit set by mouth diameter (Dodson 1970, table 5; Dodson and Dodson 1971).
2. They are visually dependent predators (Dodson and Dodson 1971).
3. Adults grow to a relatively large size, up to 8 cm snout to vent, or about 25 cm total length, considerably larger than their prey (J. R. Bizer, pers. comm.).
4. The larger adults take snails, amphipods, and insect larvae; the smaller larvae seem to rely more on zooplankton (Dodson and Dodson 1971).
5. They select *Chaoborus* over crustaceans (Dodson and Dodson 1971) and cladocerans over copepods of comparable or larger size (Sprules 1972).
6. They are predatory on salamanders (Wilbur 1972; Dodson 1970; Sprules 1972) and tadpoles (Dickman 1968).

Community Effects. The similar role of fishes and salamanders is obvious also from the resulting community effects, best illustrated by Dodson's 1970 study. Dodson distinguished two basic types of zooplankton communities in the small, shallow ponds in the mountains of Colorado.

The Type *A* community was characterized by: (1) larvae of *Ambystoma tigrinum*; (2) larvae of the phantom midge, *Chaoborus americanus*; (3) *Daphnia rosea*, a medium-sized daphnid; and (4) *Diaptomus coloradensis*, a herbivorous copepod. Type *B* communities were characterized by: (1) *Diaptomus coloradensis*; (2) *Diaptomus shoshone*, a predatory copepod; (3) *Daphnia pulex*, a very large-sized daphnid; and (4) *Branchinecta coloradensis*, the herbivorous fairy shrimp that reaches sizes above 10 mm.

The author made several observations about these two kinds of communities. First, Type *A* contained two common predatory species, and the herbivores were smaller than those associated with the Type *B* community. Second, of the fifteen Type *A* communities examined, thirteen were initially of the Type *B* assemblage; but once larval salamanders hatched out, the community changed to the Type *A* assemblage. Lastly, of the nine ponds that did not contain larval salamanders, Type *B* communities continued throughout the summer. Dodson was able to show from stomach-content analysis that predation by *Ambystoma* accounted for these changes. Once the salamanders hatched out of the eggs in the spring they selectively removed the largest, most visible prey, *Branchinecta* and *Daphnia pulex*, which were subsequently replaced in the Type *A* communities by smaller herbivores, mostly *Daphnia rosea* but also by the even smaller *Ceriodaphnia* in some ponds. Thus, the two different types of association characterized by different-sized prey were reminiscent of those found by Brooks and Dodson (1965) for the New England lakes, with the difference that in the Colorado alpine ponds salamanders, rather than alewives, were the agent responsible.

Another change in the community as a result of this gape-limited predation by *Ambystoma* was the common occurrence of *Chaoborus*. Although Dodson always found *Chaoborus* larvae in these ponds initially, only in those containing *Ambystoma* did recruitment occur, suggesting some basic advantage resulting from the salamanders. Dodson showed that a major effect of *Ambystoma*, in shifting the community from one dominated by large herbivores to one in which several smaller species predominated, was the increase in food resources for *Chaoborus*, which feed on smaller-sized prey. This shift from a community dominated by one predator (*Diaptomus shoshone* in Type *B*) to one in which two predators attain numerical importance (*Ambystoma* and *Chaoborus* in Type *A*), each feeding on a different-sized fraction of the prey population, was termed "complementary predation." This was not the creation of a new feeding niche—*D. shoshone* also feeds on smaller crustacean prey in Type *B*—or an increase in the number of predatory species as is sometimes inferred. It was a shift of predator dominance, with *Chaoborus* apparently able to replace *D. shoshone*.

This complementary relationship of predators appears in the Connecticut alewife lakes (Brooks and Dodson 1965), with the introduction of new piscine predators to a lake system (Galbraith 1967), as well as in the actions of *Ambystoma* in Colorado ponds. In all cases, the result is a community affected more nearly equally by two types of predators, one gape-limited (fish or salamanders), the other size-dependent (copepod or phantom midge). The similarity in community effects from salamander predators provides further support for the idea that amphibians are the same type of gape-limited predators as fish and are found in an identical niche otherwise occupied by fish.

Anurans (Frogs and Toads). Although some tadpoles can be carnivorous (Heyer, McDiarmid, and Weigmann 1975), none have ever been found to feed on zooplankton. Instead, most anuran young are generalized suspension feeders on phytoplankton or periphyton (attached algae). For example, studies of tadpoles of *Rana aurora* (the red-legged frog) have found them to feed almost exclusively on pond periphyton (Dickman 1968).

The absence of the evolution of zooplankton particle-selection by anurans is most likely a result of their inability to colonize successfully either large, permanent water bodies where fish could cause their demise or smaller ponds where predacious salamanders could hamper their existence. Calef (1973) demonstrated that *Taricha granulosa*, the pacific newt (which probably escapes fish predation because of its effective anti-predator device, its infamous level of toxicity), is predatory on tadpoles. He found that *Taricha* predation could account for all the tadpole mortality in the populations of *Rana aurora* that he was studying. There are known examples of a tadpole's escaping predation by its distastefulness or toxicity: the jet-black tadpoles of the large toad *Bufo americanus* (Voris and Bacon 1966) and also the Caribbean *Bufo marinus* (M. Robinson, pers. comm.) are avoided by fish predators because of their distastefulness. However, other anuran larvae are unable to survive in water bodies that contain predators, namely, fish or salamanders. This leaves only the very small and temporary ponds and puddles for colonization. In these, phytoplankton blooms occur more commonly and dependably than zooplankton. Given that phytoplankton is the most commonly available food, it is not surprising that the majority of tadpoles are adapted strictly for filter feeding (Wassersug 1975) and do not possess the long, slender body characteristic of zooplanktivores.

Other Gape-Limited Predators

In the absence of fishes or salamanders, we may still find gape-limited predators in a pond in the presence of some of the diving insects, such as members of the hemipteran families, Notonectidae (water striders), Bellostomatidae (water scorpions), Corixidae (diving bugs) (Anderson 1970), and others. Although not strictly limited by "gape," because they possess piercing and sucking mouthparts, they are included tentatively in this category because the existing scant information suggests they have the typical Class I electivity curve of figure 1. It has been suggested that the notonectid *Anisops* relies on vision to prey on daphnids (O'Brien and Vinyard 1978). Members of the genus *Notonecta*, which swim immediately under the water surface, detect prey by vibration receptors and actually inject a toxin into the captured prey (Edwards 1963). Some of the others can dive to a depth of several meters and spear zooplankton prey with their hypodermic-shaped proboscis. Although they do not have the range of a free-swimming salamander or fish, these insects are quite capable of exerting a significant predation effect in the top water levels or, in the case of a shallow pond, throughout the entire water body (e.g., Marion Lake, British Columbia, Canada; K. Hyatt, pers. comm.). These insects can occur as the dominant gape-limited predators in high altitude ponds if the ponds are inaccessible to fish and have a temperature regime unfavorable to amphibians. For instance, in the Colorado Rockies, populations of *Ambystoma* are not found to be successful at breeding much above 11,000 feet because of the severely restricted length of the growing season and reduced pond temperatures (Sexton and Bizer 1978). These ponds will not normally have active fish populations, and, consequently, this niche will be open, thus allowing insects to be the dominant predators.

As a final note, one of the most unexpected Class I predators may be the red phalarope, *Phalaropus fulicarius*, a bird that inhabits Alaskan temporary ponds, apparently feeding by visual, particle selection on copepods (Dodson and J. Erckman, pers. comm.).

4. Class II: Size-Dependent Predators

The electivity curve of the second class of predators, the size-dependent predators (SDP), is bell-shaped (fig. 17). These predators, which include only invertebrates, do not swallow their prey whole but ingest it in small bites or actually employ sucking mouthparts to "sip" their prey. In contrast to the gape-limited predators, size-dependent predators rely heavily on tactile manipulation of the prey, so that prey size is critical not so much for ingestion but for prey capture.

Although a great number of studies have provided considerable insight into the nature of vertebrate planktivores and their effects on zooplankton community structure, relatively little attention had been given other types of predators. Only recently has it been realized that invertebrate predators dominate in the absence of gape-limited predators and are capable of exerting profound effects on the prey assemblage, a conclusion reached concurrently by several investigators (Dodson 1974*a*; Zaret 1975; Kerfoot 1977*b*). The bell-shaped electivity graph with two gradually sloped sides has been documented in all of the major SDP groups including copepods, cladocerans, aquatic dipteran larvae, and rotifers. Furthermore, the two sides of the graph result from two distinct and separate causes. As with gape-limited predators, the left-hand tail results from an increased electivity of the predator from small-sized to large-sized prey. The right-hand tail is due specifically to the increasingly difficult problems of capturing and handling prey as they become larger and more capable of escaping. The effects on the prey species are distinctly different from those associated with the GLP curve; Class II predators are truly size-dependent predators. (The term "handling" as used here should not be confused with

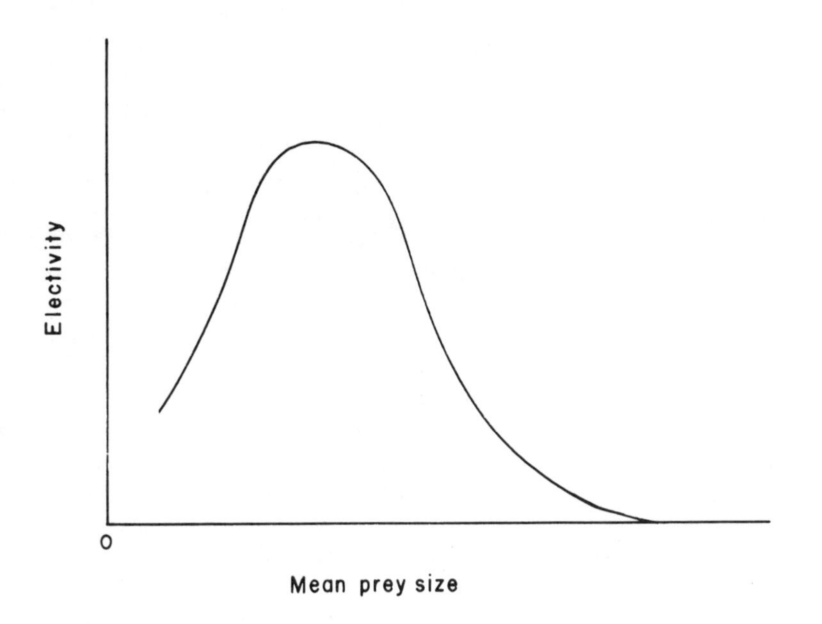

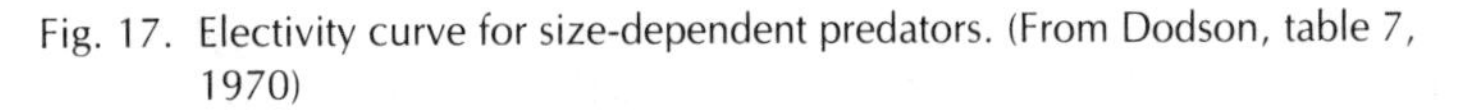

Fig. 17. Electivity curve for size-dependent predators. (From Dodson, table 7, 1970)

Holling's "handling time," which refers to the time needed to pursue, capture, and eat a prey, thus setting an upper limit to the number of prey that can be consumed in a given time. "Handling" as used here refers to the ability of the predator to physically manipulate the prey to obtain a meal and sets a limit on the size of prey the predator can subdue and consume.) The main groups of size-dependent predators associated with lake communities are described in the following section.

ENTOMOSTRACAN CRUSTACEA

Copepoda

Of the three important freshwater copepod orders, the Cyclopoida and Calanoida include important limnetic predators (the herbivorous Harpacticoida being restricted to littoral zones). Cyclopoids, especially members of the genus *Cyclops*, but also *Macrocyclops*, *Acanthocyclops*, and *Mesocyclops* (Fryer 1957), have been long recognized as primarily raptorial predators on other crustaceans, breaking off and swallowing the animal prey bite by bite (Strickler and Twombly 1975). Other cyclopoid species are herbivores (almost exclusively the genus *Eucyclops*; Fryer 1957), biting off pieces of filamentous algae in much the same way as their nonvegetarian relatives, but these are in the minority for this group. Calanoid copepods, on the other hand, were originally considered to be strictly filter-feeders (Pennak 1953), although it was suggested that some might be omnivorous. Main (1962) demonstrated that *Epischura lacustris* can be predatory on small crustaceans such as the cladoceran *Bosmina*. Thus, some of the larger calanoid copepods—*Epischura* or the common European equivalent *Heterocope*, which has been suspected of predatory behavior for a long time (Burckhardt 1944)—attain sizes of up to several millimeters and are important lake predators. This has been shown also for such other calanoids as members of the genus *Diaptomus*, including *D. shoshone, D. arcticus*, and *D. nevadensis*. Anderson (1970) has demonstrated from feeding experiments that *D. shoshone* prey upon small crustaceans at a rate faster than that of the rapacious *Cyclops vernalis*. Similar distinctions can be made for pelagic marine species (Wickstead 1962).

Prey Detection. Although they do possess eyes, copepods depend not on photoreception for detecting other animals but rather on hydrodynamic perception (Strickler 1975), coupled with olfactory sensing for sexual (Katona 1973) and other activities (Strickler and Twombly 1975). Electron microscope examination has revealed minute setae on their first antennae (see fig. 2) that act as mechanoreceptors capable of detecting gravitational and inertial forces from water disturbances (Strickler and Bal 1973). The presence of these setae explains why the calanoid copepod *Diaptomus shoshone* feeds almost as efficiently in the dark as in the light (Anderson 1967).

Locomotion. As nonvisual predators, predatory copepods must locate their prey by swimming. The frequency of copepod contact with prey (a random event) is determined by the distance from the prey and the distance traveled per darting movement of the copepod, the latter proportional to copepod body-size (in cyclopoids at least) measured as a cepahlothorax length (Rosenthal 1972). This information is most conveniently expressed as a Reynolds number, which relates the importance of inertia and viscosity. The equation for the Reynolds number is

$$R = \frac{PUL}{u}$$

where P is the density of the medium, u is the viscosity of the medium, U is the velocity of the moving object, and L is the length of the object. When the Reynolds number is less than 1,000, the flow around the object is considered nonturbulent, or laminar, which means that it is nondeterminant or random.

Cyclopoid copepods stroke with their two long antennae in a darting motion. Taking into consideration the viscosity of water, this normal hopping or skipping motion—often referred to as "hop and sink" because cyclopoids pause after each hop and sink down because of gravity—includes a rate of one antennal power stroke per second, attaining a mean speed of 0.1 to 0.5 cm per second and a maximum Reynolds number of 50. During its escape response, however, cyclopoids may reach 120 power strokes per second, a maximum speed of 35 cm per second, and a Reynolds number approaching 500. This is their maximum range (R. E. Zaret, in press). In contrast, a 5 cm fish moving 10 cm in water will have a Reynolds number of approximately 5,000. At low Reynolds numbers (<100) swimming becomes difficult because the slower the object moves, the more it is affected by objects in its natural environment. The closer a small animal is to a stationary object, the more it is slowed. At Reynolds numbers of the higher range (500), the streamlining of copepods becomes important because it reduces drag.

In contrast to cyclopoids, calanoid copepods never attain high Reynolds numbers (Strickler 1975). Rather than employ their antennae, calanoids use their mouthparts to gently propel themselves. Their locomotion is characterized by gliding back and forth in the water column.

Evolutionary Comparisons. Some interesting evolutionary questions arise concerning the relationship of these two copepod groups and their feeding modes. Although there is general agreement that cyclopods are more primitive and calanoids more advanced, there is a difference of opinion about whether filter feeding or raptorial feeding is more recently evolved. Tiegs and Manton (1957) feel the filter-feeding mode is more primitive but was secondarily lost in the cyclopoids. Fryer (1957) concludes the reverse. In any case, the exact method of predation in the two groups is sufficiently different that either conclusion is possible.

Copepods detect prey by hydrodynamic information, which is produced by the wakes of prey organisms. When a copepod chances upon a wake, a random event, it can determine the biomass-velocity product of the animal that just passed. If the wake is sufficiently small, the predatory copepod will either not recognize it or ignore it. If the wake is very large, the copepod immediately will initiate escape responses, and there is apparently a strong negative rheotaxis reflex in all copepod species examined (Strickler 1975). If the wake is of the appropriate size range, the predatory copepod will follow it. If it also contains the species' pheromone, the copepod will attempt sexual behavior (Katona 1973); otherwise it will seize it as prey (Strickler and Twombly 1975). For the actual capture, cyclopoid copepods follow in the prey wake and rapidly attack with one burst, stabbing the prey from behind with daggerlike mandibles. The calanoids, on the other hand, do not actually enter the wake but swim just above it, apparently able to receive mechanoreception information while remaining undetected by the potential prey below. When a predatory calanoid achieves a position above and parallel to the prey, it suddenly dives and seizes the prey with its mouthparts, primarily the maxillipeds. Thus, for cyclopoids, the attempt is to overwhelm the prey immediately; with calanoids, stealth is employed. The disturbance produced by a male calanoid swimming above is not detectable by the female swimming below; otherwise, she would immediately initiate an escape reaction and copulation would never take place. This may be one reason for the size dimorphism in many copepods in which the male is usually the smaller (and less detectable?) of the two (Wilson 1959). In *Diaptomus shoshone*, for example, larger males tend to mate with larger females (Maly 1978).

Cladoceran carapaces should present difficulties for predators trying to capture, pierce, and crush them with their mandibles, a characteristic of the cyclopoids. The characteristic dome shape of the cladoceran carapace is known for strength; stress is distributed more evenly over the entire structure. Additionally, this shape may be difficult to grasp, like a slippery bar of soap (Strickler, pers. comm.). Furthermore, at their very low Reynolds numbers (numbers for *Bosmina* range to a maximum of 3), this egglike shape of cladocerans is of no value as a means of decreasing friction and thus increasing swimming speed because particles of this size are moving within an envelope of water whose thickness is affected little by their morphology. Only at higher Reynolds numbers (close to 500) does streamlining become important in reducing drag (Strickler 1975). It is therefore likely that the cladoceran shape has evolved at least in part as a response to specific predation pressures.

The particular ecological characteristics of calanoids appear to make them ideally suited for the limnetic mode of life. They can swim continually for up to sixty seconds by means of propulsion with their mouthparts, an energetically inexpensive way to glide back and forth in the water column in search of prey. As a result, an adult *Epischura* covers considerably more area per unit time in its foraging pattern than a cyclopoid such as *Cyclops*, which uses only the sudden jerking of the hop and skip motion (Strickler and Twombley 1975). This searching pattern of calanoids is adaptive for the limnetic regions, which, often relative to the littoral zones, are characterized by low prey densities. Calanoid copepods, moreover, cautiously "stalk" their prey, which is also adaptive in the open areas of the lake where the predator must maximize its strike success; a missed prey could escape in any direction and make recapture very unlikely. Finally, a calanoid copepod does not attempt to crush its prey but manipulates it until the exposed and vulnerable areas of the valve opening are reached (Kerfoot 1977*a*), a behavior that seems especially adapted to feeding on the bivalved cladocerans, the predominant limnetic zooplankton group.

In contrast, cyclopoid copepods seem to be adaptive for the littoral zones, which are usually characterized by abundant hiding places for prey, including attached and rooted plants as well as bottom substrate. Cyclopoids are really ambush predators, remaining still and then suddenly darting toward the prey in a strong, quick burst of movement. Once the prey is detected, the predator must immediately capture it before it returns to its refuge. The actual attack is very rapid, with a premium on striking speed.

Based on these ecological arguments, it appears possible that freshwater calanoid copepods have evolved primarily for existence in the limnetic regions of lakes, being derived after the cyclopoids whose normal habitat is the littoral region. Initially, this allowed calanoid filter-feeders to exploit the great quantities of algae. Secondarily, the raptorial feeding mode may have evolved in the calanoid group (just as it may have evolved secondarily in the cyclopoids) to allow feeding on the abundant zooplankton, primarily cladocerans, in the limnetic zones. It is true that cyclopoid copepods are often encountered in limnetic lake regions, but in these cases it is often the subadult copepodids that are found rather than the adults, which are normally the dominants in the littoral zone. Calanoids are the open-water dominants.

Community Effects. In spite of the distinct differences in feeding methods between the two dominant freshwater copepod groups, the effect of their attacks on a prey population is identical. Because the actual feeding mechanism involves the grasping of the potential prey, handling becomes the critical factor. The size of the prey is limited by the size of the predator's mouth appendages, especially its maxillipeds (Anderson 1967). Smaller prey are more easily handled and more readily captured and eaten. This means that although the electivity curve for such predators as copepods increases with size up to a distinct range, it drops off according to the size of mouthparts as prey become more difficult to capture because of escape responses or handling problems. This relationship of electivity and prey size (fig. 17) has been shown for the cyclopoids *Mesocyclops edax*, *Cyclops vicinus*, and *Acanthocyclops vernalis* feeding on the cladoceran *Ceriodaphnia* sp. (Brandl and Fernando 1975); *Acanthocyclops vividis* feeding on the cladoceran *Simocephalus vetulus* (Smyly 1970); and has been suggested for other copepod species (Anderson 1970). In addition, Confer (1971) showed that adult *Mesocyclops edax* preferred immatures of the calanoid *Diaptomus floridanus* to adults; and Dodson (1974*b*) demonstrated the same curve for the calanoid *Diaptomus shoshone* feeding on *Daphnia* and *Ceriodaphnia*.

The same result has been demonstrated in laboratory work with *Epischura* feeding on different cladocerans. Kerfoot (1977*a*) determined that the increasing death rate of the small cladoceran *Bosmina longirostris* in Lake Washington was correlated with the abundance of *Epischura nevadensis*. To test the effect of predation by this calanoid copepod, he ran a laboratory feeding experiment in 10-liter aquaria to which were introduced 100 immature instars of each of three different prey species from the lake: *Bosmina* (body-size 0.2–0.3 mm); *Ceriodaphnia* sp. (body-size 0.3–0.4 mm); and *Daphnia ambigua* (body-size 0.3–0.4 mm). Eight adult *Epischura* were introduced to each experimental aquarium, and aquaria with prey species only were used as controls. The experiment was run for three days, with the number of remaining prey counted at one-day intervals. Figure 18 presents the averaged results of two replicates of the experiment. Each bar represents the cumulative number of animals removed from the experimental aquaria minus the number that died in the controls. The study showed a significant decrease for both *Bosmina* and *Ceriodaphnia*, with *Bosmina* preferred most and *Daphnia* least, indicating the selection by the copepod predators of the smaller prey sizes.

Copepod predators will attack neither very large nor very small prey. For those prey captured, the effect of size again becomes important, because copepods are not efficient in the handling of larger objects even after

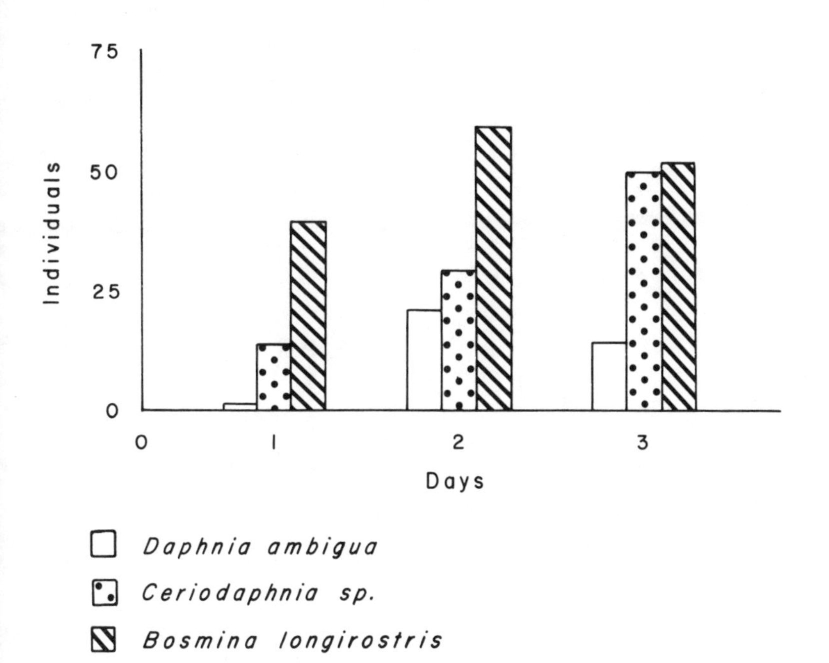

Fig. 18. Effect of prey size on predator selectivity. Cumulative mortality of three different-sized cladoceran prey in the presence of *Epischura lacustris* over a three-day experiment. (From Kerfoot 1976).

they have captured them. The result is the distinctly bell-shaped electivity curve that characterizes these size-dependent predators, the gradually decreasing right-hand side distinguishing them from Class I predators.

COPEPOD MODEL

In figure 19, another schematic model of predation, predatory copepods are the example chosen as the size-dependent predator, for the simple reason that the greatest amount of current knowledge exists for this group. Again, the model does not include the effect of higher-order interaction from predators (this will be taken up later), and it assumes short time intervals throughout. The model can be generalized to other size-dependent aquatic predators, some of which, however, require light to detect prey, which copepods apparently do not.

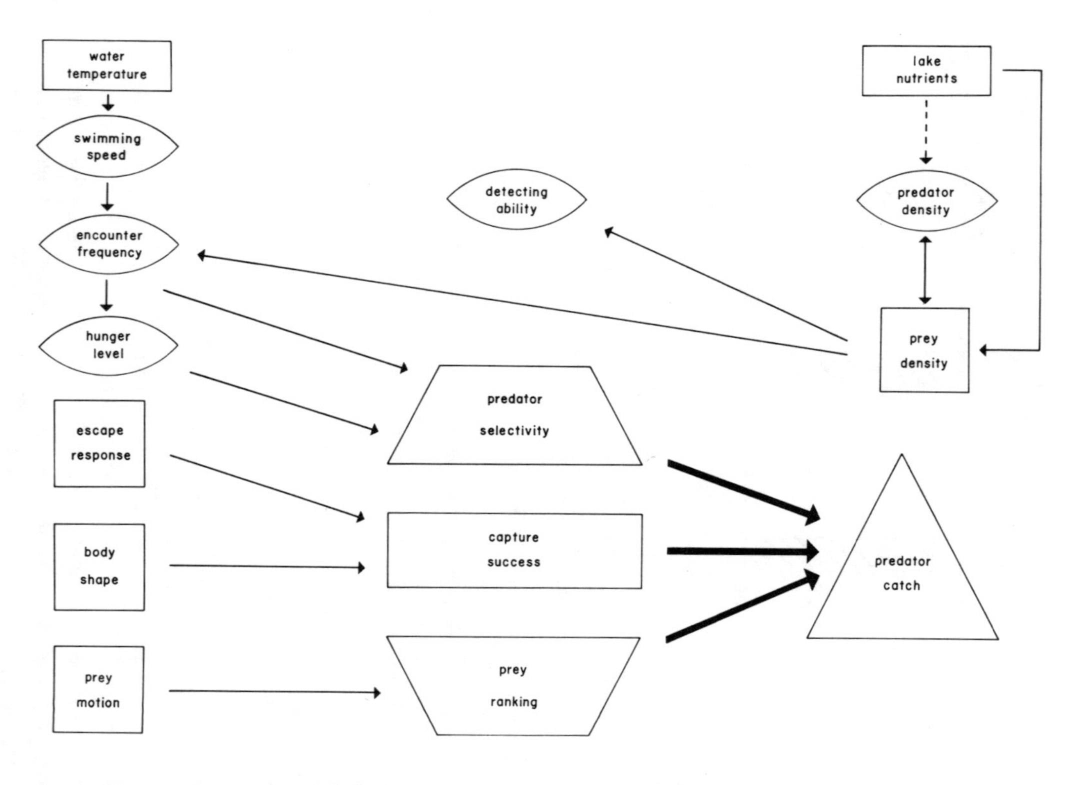

Fig. 19. Copepod model (the bold lines identify the major determinants of predator catch; the regular solid lines signify proven relationships; the dotted line signifies a questionable relationship).

Physical Environment Characteristics

Water temperature has been shown to affect predation rates in other invertebrate predators (Fedorenko 1975*b*), presumably by affecting swimming speed and thus encounter frequency, and this effect is assumed to apply also to predatory copepods. The only other important aspect of the physical environment is the level of lake nutrients, which will control prey density and, possibly, predator density.

Predator Characteristics

There are major differences between this model and that for the gape-limited predator. First, the detecting ability of this predator is not affected significantly by any aquatic characteristic of the environment—any water movements would be on too large a scale to affect the ability of the predator to detect that of the prey. Rather, it is the density of the predator itself—and perhaps the density of the prey—that bears upon detecting ability. This surprising conclusion arises from the fact that a copepod, upon detecting a large water movement, such as one produced by an animal as large as itself, will exhibit a rapid avoidance response and thus interrupt time spent on prey search. Thus, with higher predator densities or large-sized prey, we would expect frequent interruptions and a decrease in search time by the copepods, a strong density-dependent effect.

Encounter frequency is affected by only two factors: the swimming speed of the copepod and the density of the prey. The swimming speed of the copepod determines how much space will be covered in each darting movement of cyclopoids or gliding movement of calanoids during prey search. As mentioned previously, in cyclopoids this is proportional to body-length. The most important factor, however, is the actual prey density. Because swimming speed, as a function of predator body-size and water temperature, cannot be changed over time, it is through prey density that the prey encounter frequency is regulated in natural situations.

The population density of prey plays a highly critical role for size-dependent predators, who rely on random encounters with prey wake for detecting prey motion. Gape-limited predators can cover much greater distances and visually locate high densities of prey from greater distances. However, some size-dependent predators also are able to see their prey. Invertebrate predators, therefore, should be much more sensitive to low-density prey levels; and there is some suggestion that certain *Chaoborus* species, for instance, are much more effective at low prey densities than others (Fedorenko 1975*a*, 1975*b*). This observation has been largely ignored, but its importance will become clear later on in this book.

Encounter frequency is the most important determinant of hunger level. As with fish, selectivity is determined solely through encounter frequency and hunger level.

Prey Characteristics

The important prey characteristics are prey motion, escape response, and prey shape. Prey motion is the sole source of information transmitted to copepod predators and determines completely the prey ranking. Given this ranking, the other two prey characteristics, escape response and prey shape, determine the capture success of those prey pursued and attacked by the copepod. Included in prey shape is total body-size, which may affect also prey motion, as well as those morphological structures that interfere with capture frequency. It is the interaction of selectivity, capture success, and prey ranking that results in the predator catch.

Thus, the copepod model differs from the model for vertebrate planktivores in three important ways: (1) the relatively reduced importance of physical environment characteristics, (2) the dependence on random predator-prey interaction, and (3) following from (2), the overriding importance of prey density in affecting predatory copepod abundances, especially critical at low prey densities.

Support for this last conclusion comes from recent mathematical models of invertebrate predators. While searching, the encounter frequency of a predator such as a praying mantid is considered to be represented as a Poisson stochastic process, producing an expected number of encounters per unit time directly proportional to prey density (Holling 1966; Charnov 1973).

OTHER SIZE-DEPENDENT PREDATORS

Anostraca

Fairy shrimp (fig. 20), relatively primitive, shrimplike entomostracan crustaceans which swim upside-down, most often inhabit temporary ponds and pools. Fewer than thirty species occur in North America, commonly in high mountain alpine ponds. No strictly marine representatives exist, although the best-known species is the brine shrimp *Artemia salina*, which lives in saline water bodies, including the Great Salt Lake in Utah. Anostraca are not able to survive for long in ponds containing fish, but under

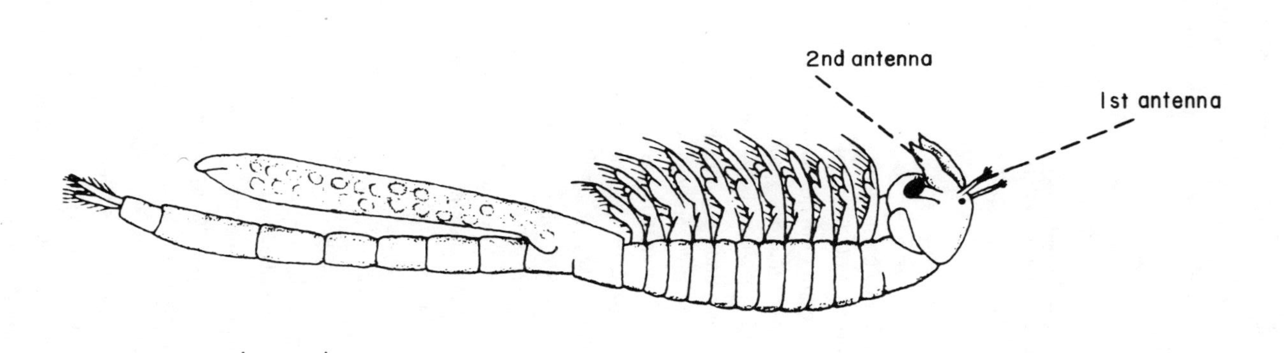

Fig. 20. The predatory anostracan *Branchinecta gigas*

certain conditions they can survive in those with predatory amphibians or insects (Dexter 1959; Dodson 1970; Sprules 1972), which have lower GLP feeding efficiencies.

Most Anostracans are filter- and detritus-feeders, but the family Branchinectidae, of which there is only one genus with seven species, includes the predatory *Branchinecta gigas*. This species can attain body-lengths of up to 100 mm and will take cladocerans, copepods, and the anostracan *B. machini* (Anderson 1970). The latter is its common prey in natural habitats (White, Fabris, and Hartland-Rowe 1969; Daborn 1975).

Cladocera

The Cladocera represent a group of crustaceans that has evolved specifically for fresh water. In distinct ecological contrast, copepods are the dominant marine crustaceans, being a major source of food for oceanic fish, and have invaded freshwater situations secondarily. The most unique characteristic of the Cladocera is the possession of a bivalve carapace, which, with few exceptions, covers and protects the body and into which all appendages other than the antennae can usually be withdrawn (see fig. 3). Only in predatory freshwater cladocerans, namely, *Leptodora* (fig. 21a), *Polyphemus*, and *Bythotrephes* (fig. 21b), has the carapace been reduced, that is, limited to covering only the brood sac. Presumably these species, which attain lengths of up to 18 mm, gain no important advantage (e.g., protection) from this structure, as they usually are not preyed upon by each other or by such other invertebrate predators as the phantom midge, *Chaoborus* (Cummins et al. 1969). These three predatory genera are found in lakes around the world, including very large ones, such as the Great Lakes.

Polyphemus pediculus is the smallest common member of the plankton whose many-lensed eye is capable of forming a distinct image and is used to locate prey (Brooks 1959). Studies by Bosch and Taylor (1973*a*, 1973*b*) on the closely related and similarly appearing *Podon polyphemoides*, which lives in the Chesapeake Bay estuary, indicate that it ceases feeding after sunset and is thus a light-dependent predator. (Bainbridge, 1958, found this true also for the related estuarine *Evadne nordmanni*.) *Polyphemus* captures prey and then forces them into the food chamber, grinds them up with its mandibles, and sucks in the ground-up food (Monakov 1972). Occasionally, this species has been reported as herbivorous. It has been reared on bacteria and protozoans (Butorina 1971).

The sole species in the genus, *Leptodora kindti* has been studied extensively. Although usually exhibiting low numbers, it sometimes attains

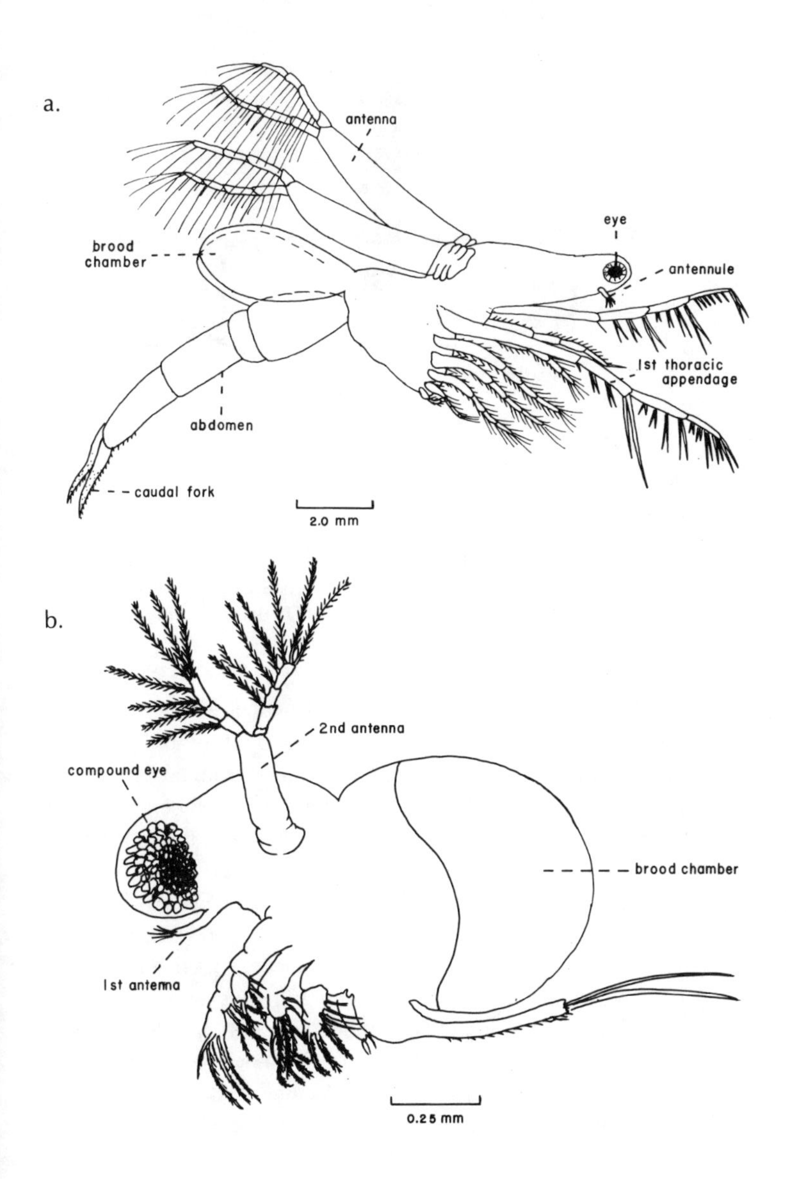

Fig. 21 a. The cladoceran *Leptodora kindti*. b. The cladoceran *Bythotrephes longimanus*

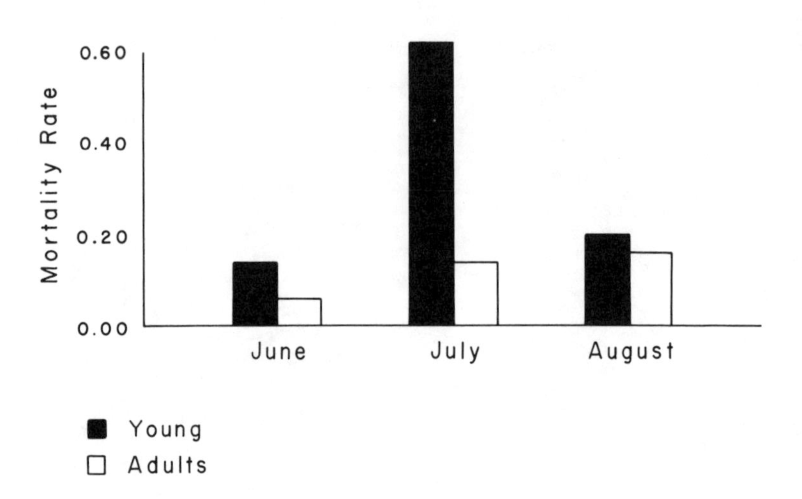

Fig. 22. Effect of prey size on predator selectivity. Average death rates for young and adult *Daphnia hyalina* during months of predation by *Leptodora* and *Bythotrephes* in Lake Maggiore, Italy. (From De Bernardi and Giussani 1975)

extremely high densities, up to several hundred animals per liter (Cummins et al. 1969; Hillbricht-Ilkowska and Karabin 1970). The adult is almost exclusively predacious, capturing prey in the "basket" formed by its swimming legs and then piercing the prey with the sharp protuberances of its pointed mandibles (Sebestyén 1931) to suck out juices. Adults often reach nearly 20 mm and can take prey up to 1.5 mm (Sebestyén 1960). On occasion the prey can be swallowed whole, integument and all (pers. obs.).

Predatory cladocerans are typical size-dependent predators. In their study of Lake Maggiore in Italy, De Bernardi and Giussani (1975) calculated prey death rates and found that *Bythotrephes* and *Leptodora* selected the smaller, immature members of *Daphnia hyalina* over larger individuals (fig. 22). A similar conclusion for *Leptodora* feeding on daphnids has been suggested from other field studies (Hall 1964) as well as laboratory feeding experiments using the small cladocerans *Bosmina* and *Ceriodaphnia* as prey (Mordukhai-Boltovskaia 1958).

Community Effects. The impact of size-dependent cladoceran predators on lake species composition has been illustrated by several field studies. De Bernardi and Giussani (1975) showed that several *Coregonus* species preferentially removed *Bythotrephes longimanus* and *Leptodora kindti* over the smaller *Daphnia hyalina*, consuming the largest individuals of each species. Once the predatory cladocerans were reduced, however, the fish species turned their attentions to the daphnids. It had been shown that these fish limited the spring densities of *Daphnia* populations by removing selectively the adult fertile instars, thus controlling population growth. Normally in Lake Maggiore, the daphnid population remains low for the remainder of the year, and previous studies had concluded that fish were responsible for year-round control of these prey. Although a large population of the coregonids and the usual low level of *Daphnia* were observed in 1972, in 1973 the planktivore populations were severely diminished. In 1973 only, during the spring months of May and June, the populations of daphnids soared to densities above 10,000 per liter, compared with levels of approximately 200 per liter in 1972, thus demonstrating the effects of the absence of *Coregonus* populations. Starting in late June and early July 1973, however, the daphnid population suddenly returned to levels associated with previous years, coinciding with the appearance in the lake of the predatory cladocerans *Leptodora* and *Bythotrephes*. Through calculations of mortality, the authors were able to show that the summer decline in daphnids was due to removal of young instars by these two size-dependent predators (fig. 22).

The conclusions from this fortuitous manipulation of fish densities were that the spring abundances of daphnids were controlled by fish feeding on the adults, but the summer collapse was affected by the invertebrate predators feeding on the young instars. The adult daphnids are not as strongly affected by fishes during late summer because at that time the planktivores are larger and less planktivorous and thus have little impact on the daphnids.

Other studies also have demonstrated the effects of predation by size-dependent predators. Hall (1964), from work on Base Line Lake, Michigan, was able to show that the population of large daphnids, *D. pulex*, was controlled by fish predation—the coregonid known as cisco, *Leucichthys artedi*, a Salmonidae, and the black crappie, *Pomoxis nigromaculatus*, a Centrarchidae. The smaller daphnid species *D. galeata mendotae*, also present, was ignored by fish but preyed upon exclusively by *L. kindti* and *Chaoborus*. In a study by Wright (1965), the author concluded that the summer decline of *D. schodleri* in Canyon Reservoir, Washington, was due to predation by *Leptodora*.

The agreement of studies from three different localities concerning the impact of invertebrate predators is convincing evidence that these size-dependent predators are fully capable of controlling the development of prey populations in natural conditions by the selective removal of the smallest instars of the population. These studies also confirm the earlier conclusions of Smith (1963) on the ability of size-dependent predators to affect prey densities.

Malacostraca Crustaceans

Mysidacea. The mysids, or opossum shrimps, are regarded as the most primitive of the freshwater malacostracan orders (Chace et al. 1959). Their common name derives from their unusual means of parental care. Eggs are brooded and young are sheltered in a brood pouch, or marsupium, until they become independent. Of the three North American species, the most important by far is *Mysis relicta*, considered to have been carried from its brackish-water habitats by Pleistocene glaciers and left behind as the glaciers retreated. *Mysis* is present especially in such large water bodies as the Great Lakes of the United States and the Finger Lakes in New York State; it is present also in northern Europe. The feeding mode of *Mysis* can be herbivorous, detritivorous, or predatory (Bowers and Grossnickle 1978). In some lakes it feeds on benthic detritus by day and then actively migrates upward at night (Beeton 1960) to feed on cladocerans, for example, *Daphnia pulex* (Lasenby and Langford 1973). Richards et al. (1975) report that the introduction of *Mysis relicta* to Lake Tahoe may have been the major cause of a decline in the formerly abundant *Daphnia* populations. Its overall importance in controlling zooplankton populations has only recently been evaluated (Goldman et al. 1979, studies in Lake Tahoe; Edmondson, Murtaugh, and associates for *Neomysis* in Lake Washington).

Amphipoda. Amphipods are usually considered to be benthic crustaceans living on detritus, such as the species *Hyalella azteca*, which has a widespread distribution (see Strong 1972). There are also limnetic species predatory on zooplankton. Kozhov (1963) has described the pelagic amphipod *Macrohectopus* in Lake Baikal, Russia, as an actively migrating species, feeding heavily on crustacean zooplankton and itself being a most important food source for fishes. Recent studies on *Gammarus*, a common temperate zone amphipod, have shown at least one species to be limnetic at times and a significant predator on *Chaoborus* and crustaceans (Anderson and Raasveldt 1974). These authors suggest that *G. lacustris* may be able to prevent the establishment of *Chaoborus* populations in water bodies not containing fishes, the latter condition normally sufficient to extinguish limnetic *Gammarus* species. Laboratory feeding experiments have demonstrated a typical size-dependent feeding curve in *Gammarus* feeding on crustaceans (Anderson and Raasveldt 1974).

INSECTA

Chaoboridae. By far the most important lake predator among the insects is the phantom midge, *Chaoborus*, of which there are approximately eight North American species (James 1959). This predator, which can be very abundant in lakes (with up to two or more species present concurrently), has been studied extensively (LaRow 1970; Dodson 1970; Fedorenko 1975*a*, 1975*b*; Fedorenko and Swift 1972; Swift and Fedorenko 1973, 1975; Lewis 1975; Pastorok 1978; von Ende 1979). The aquatic larvae are highly predacious, especially in their later instars, with three to four larval stages before their emergence as adults. Some species have a one-year cycle; others have cycles of two years before maturity. In general, *Chaoborus* remains at the bottom or within the lake sediment during the day and ascends at night to feed, but this pattern is not always true and they may feed diurnally just above the bottom (Parma 1971). Possibly related to this behavior, *Chaoborus* is the only freshwater invertebrate with well-developed hydrostatic organs (Hrbáček 1977).

Although the *Chaoborus* diet is influenced strongly by local food availability (Fedorenko 1975*b*), this predator definitely selects prey according to size. Fedorenko (1975*a*) found, from gut analysis, that second and third instars of *Chaoborus americanus* and *C. trivittatus* in Eunice Lake, British Columbia, experienced difficulty in eating the largest available *Diaptomus kenai* individuals. For daphnid prey, the upper size limit was 2.2–2.6 mm for *C. trivittatus* and 2.2 mm for the smaller *C. americanus*, although the *Chaoborus* were more likely to actually avoid these large-sized *Daphnia* than to strike at them (Swift and Fedorenko 1975). The authors found that the total time for predation increased in direct proportion to increasing prey size (fig. 23). (They use the term "handling time," defined as the time from predator-prey contact until the prey is ingested past the posterior margin of the larval head capsule.) Daphnids and copepods below 0.6 mm were too small to be captured, so the *Chaoborus* electivity

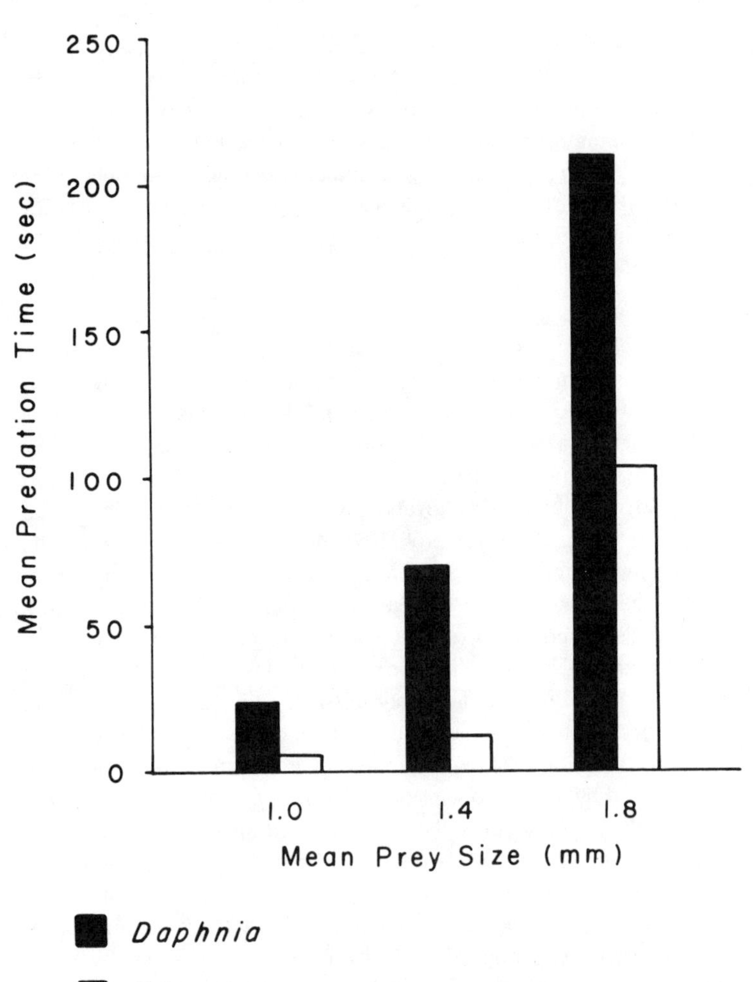

Fig. 23. Relationship of mean predation time required for ingestion of different-sized crustacean prey. (From Swift and Fedorenko 1975)

curve is bell-shaped, typical of size-dependent predators. Figure 17 comes from Dodson's electivity data (table 7, 1970) from his study on *C. americanus* in Colorado alpine ponds.

Many workers have concluded that *Chaoborus* takes copepods over similarly sized cladocerans (Main 1962; Roth 1971; Sprules 1972; Swift and Fedorenko 1975), although not all agree (Dodson 1970, 1974*b*; Lewis 1975). As support for this suggestion, Swift and Fedorenko (1975), in the previously mentioned study of the feeding habits of *Chaoborus trivittatus* and *C. americanus*, also considered "contact efficiency," by which they meant the percentage of times in which a strike by *Chaoborus* hit or held a prey and resulted in the prey's ingestion by the predator. For prey above 1 mm the contact efficiency of both *Chaoborus* species was higher for copepods than for cladocerans. This finding, together with the previous observation that total predation time is significantly higher for cladocerans (see fig. 23), strongly suggests that handling problems after predator contact result in a greater likelihood of cladocerans escaping than copepods. In fact, Swift and Fedorenko (1975) found that *Chaoborus* ingested the calanoid copepod *Diaptomus* up to seven times faster than *Daphnia* of identical size. The difference, especially in handling time, implies that the specific shapes of these crustaceans are reflected in predator ingestion rates and electivities. Other laboratory studies with *Chaoborus* have also concluded that its electivity is higher for copepods than for cladocerans of similar size (Sprules 1972). It has been suggested that the cigar-shaped body of copepods makes them easier to swallow than cladocerans (Roth 1971), although *Chaoborus* may actually strike more frequently at cladocerans because of their more conspicuous motion (Pastorok 1978).

Little is known about the exact feeding methods of *Chaoborus*. Because *Chaoborus* is perfectly capable of capturing prey in complete darkness (Dühr 1955), it has been suggested that it is capable of feeding both day and night (Goldspink and Scott 1971; Lewis 1975). Its known nocturnal activity also implies that it is not dependent on light for feeding. *Chaoborus* searches for prey by swimming in brief gliding bursts and probably responds to water movements analogous to the response of copepods. Any invertebrate using nonvisual cues for the detection of zooplankton will have evolved some kind of mechanoreceptors similar to those found in copepods. The major difference between *Chaoborus* and copepods as predators is that the former swallows prey whole, which has implications for both feeding and assimilation efficiency of the *Chaoborus* as well as its ability to consume different-shaped prey.

The effect of *Chaoborus* predation on zooplankton populations can be highly significant. Dodson (1972) calculated that *Chaoborus* accounted for up to 90 percent of the mortality of a *D. rosea* population, although it is clear he badly underestimated the contribution of such other predators as copepods (see Dodson 1974*b*). Fedorenko (1975*b*) estimated that *Chaoborus* could remove almost 10 percent of the available zooplankton per day.

In the marine environment the numerous chaetognaths, known commonly as arrowworms, are remarkably similar to *Chaoborus* in their predatory nature, dependence on tactile stimuli and mechanoreception for prey detection, and other aspects of their ecology (Reeve 1970; Feigenbaum and Reeve 1977). Their role in pelagic communities is still relatively unknown.

Arachnida. Some of the water mites have been seen to capture diaptomid copepods (Anderson 1971) and daphnids (Dodson 1972). In Madden Lake, Panama, the mite *Piona limnetica* may be the first truly limnetic species of arachnid, being morphologically adapted for swimming (i.e., its long swimming hairs are flattened). Gliwicz and Biesiadka (1975) found densities of up to 2,000 individuals/m^2 for nymphs and 150/m^2 for the imagenes. *Piona* selectively avoids rotifers and prefers cladocerans, especially *Diaphanosoma* and *Bosmina*, over copepods. *Piona*'s consumption rate was estimated at 10 to 20 individuals per day, at the maximum predator density, which amounted to a highly significant 50 percent of the standing crop of cladocerans per week.

Odonata. The aquatic larvae of dragonflies and damselflies can be important predators on zooplankton (Johnson 1973). Even though some species are adapted somewhat for swimming, most are ambush predators and restricted to the littoral regions, so that only in temporary or small ponds will they affect limnetic prey. Odonates have well-developed compound eyes and use these to capture prey. Studies of damselflies, however, have actually found that prey detection is dependent on tactile cues (Johnson, Akre, and Crowley 1975). Preliminary unpublished experiments suggest that these predators exhibit a size-dependent electivity curve.

Rotifera. Whereas most rotifers are filter-feeding herbivores, some species are predatory, although these are usually considered omnivorous (Edmondson 1959). Predatory genera include *Synchaeta, Trichocerca, Polyarthra, Ploesoma,* and *Dicranophorus*; by far the most important are the members of the family Asplanchnidae, some of which attain sizes of up to 1.5 mm and more (Gilbert 1973*a*). This last family is the only rotifer group with what is termed an "incudate" feeding apparatus, a pincerlike affair extended outside the mouth to grasp prey, which are then drawn back into the mouth. Asplanchnidae possess an extensible sac on the dorsal wall of the pharynx that allows them to ingest rather large-sized prey.

Several studies of *Asplanchna* have shown this predator to display a typical size-dependent electivity curve, based on an inability to seize very small prey with the incudate feeding apparatus and increasingly difficult handling problems with large prey (Gilbert 1967; Pourriot 1974). This suggests that such predatory rotifers as *Asplanchna* will feed most successfully on rotifer prey, which are usually smaller than 0.3 mm, and should feed on crustaceans less frequently because such prey are usually larger than 0.3 mm and present handling problems.

Predatory rotifers, including *Asplanchna*, do seem to prefer rotifer prey when the latter are abundant, but they will switch to crustacean prey including copepods and cladocerans if the rotifer density decreases (Hurlbert et al. 1972*a*).

The density of the voracious *Asplanchna* has been recorded as high as 200 animals per liter of water (Hurlbert et al. 1972*a*), so that their effect on coexisting rotifer populations can be considerable.

5. Prey Evolutionary Responses

Given the specific selective forces exerted by the two classes of freshwater zooplankton predators, it remains to discuss the successful prey responses over brief and over evolutionary time-scales. I will review the type of sensory information used by each class of predator for prey detection and the evolution of the prey population in response to each of these selective pressures.

GAPE-LIMITED PREDATORS

LIGHT

Gape-limited—fishes, the most important group, followed by aquatic salamanders and insects—are particle selectors. As such, they rely heavily on light for prey detection and capture (Blaxter 1966; Ivlev 1961; Dodson and Dodson 1971; Zaret 1971, 1975; Suffern 1973). Prey individuals that can live successfully in areas of reduced light will have a much better chance of escaping this predator; thus we would expect an increase in light-avoidance behavior (with a heritable basis) over time.

One successful prey response is the "strategy" (i.e., evolved adaptation) of vertical migration, whereby prey exhibit a diurnal pattern of migrating to levels of low light intensity where predation pressures are reduced. Vertical migration in freshwater and marine aquatic organisms has been documented for a wide diversity of animals, primarily crustaceans, but also chaetognaths (arrowworms), coelenterates (e.g., jellyfishes), insect larvae, fish, and others (see McLaren 1963, Banse 1964, and Hutchinson 1967 for reviews). Some of the various hypotheses to explain the adaptive significance of this behavior include more efficient utilization of phytoplankton populations (Hardy 1956); increased genetic exchange of migrating organisms (David 1961); population self-regulation based on group selection (Wynne-Edwards 1962); increased energy efficiency (McLaren 1963); increased productivity (McLaren 1974); and avoidance of competition (Dumont 1972, Lane 1975). Although many theories have been developed to provide a single explanation for all vertical migration patterns, it is probable that this behavioral accommodation may provide different benefits at different times, in different organisms, depending on the particular dominant selective pressures on that population (Hutchinson 1967; Zaret and Suffern 1976).

Recent experimental evidence supports the effectiveness of vertical migration as an anti-predator mechanism (Zaret and Suffern 1976). Data from Gatun Lake, Panama, show that the numerically dominant crustacean zooplankton is the calanoid copepod *Diaptomus gatunensis* (see fig. 12) and the important planktivore, the atherinid *Melaniris chagresi* (see fig. 14). Stomach contents of the fish indicated that *Diaptomus* was rarely taken as a food item (less than 5 percent of the time), even though this copepod was the most abundant and largest prey item within the limnetic zone. In the laboratory, when presented with a choice of zooplankton prey items, *Melaniris* showed a very high feeding rate on *Diaptomus*, with its greatest electivity for the adult stage, then the copepodid stage, and a strong negative value for the small nauplii stage. *Melaniris*, therefore, can feed on the copepods and demonstrates a very high electivity for them in the laboratory but rarely consumes them in the field.

Vertical migration studies showed that during the daytime the adult copepods concentrated near the bottom of the lake but during late afternoon migrated upward. Just after dusk they were located en masse in the top several meters of water. During the night they gradually descended to the bottom. The same pattern, although less striking, was found for the copepodids, and an even weaker pattern was observed for the nauplii (fig. 24). This is the normal diurnal pattern for migrating crustaceans. Because the fish fed only in the top meter of water, and because they were restricted for feeding to a minimum light level, probably around 1 lux, with feeding activities ceasing just after dusk, the migration pattern of the copepod brought it to surface waters only when *Melaniris* was incapable of feeding. This behavior was highly successful, and *Diaptomus* was the dominant

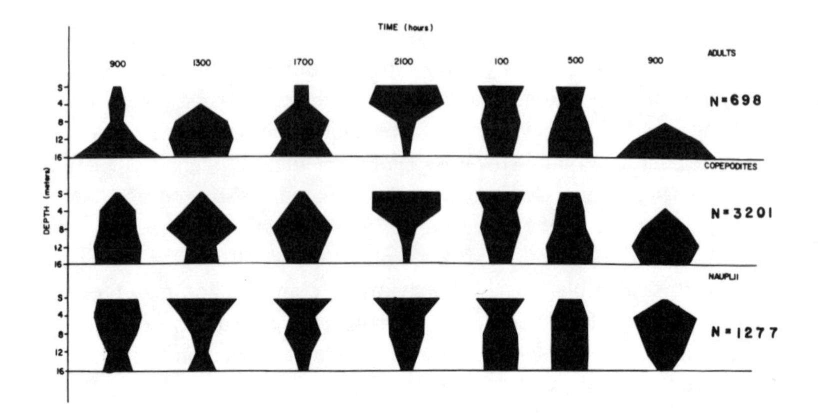

Fig. 24. Vertical migration pattern of *Diaptomus gatunensis* in Gatun Lake, Panama. (From Zaret and Suffern 1976)

lake zooplankter. Under natural selection, copepods that migrate upward when the fish are feeding will be removed (a certain fraction of the crustaceans must do this because fish do consume copepods), leaving only those individuals with the adaptive migration pattern.

A second example of this inverse correlation of potential zooplankton prey distribution with the predator's feeding intensity was found in Fuller Pond, Connecticut. In this case the cladoceran *Daphnia galeata mendotae* was shown to be an important prey item for the dominant planktivore, *Notemigonus crysoleucas* (Cyprinidae). *Notemigonus* fed along the littoral margins of the lake diurnally, migrating out into more open water only at sunset to feed on the daphnids, preferring the adults. The daphnids in this lake also showed a pronounced vertical migration pattern (fig. 25), with the main diurnal cluster of adults around the 8 to 9 m depth, rising to 3 m or less after sunset. Laboratory experiments with *Notemigonus* feeding under different light intensities demonstrated that the light levels associated with the daytime holding depth of the daphnids were in the range where planktivore feeding efficiency was very low, which suggested that the daphnids' migration pattern reduced predation and also that the patterns of the crustaceans might change considerably if the fish were not present. As with Gatun Lake, a certain number of daphnids are removed by the fish, which indicates that selection against those individuals with the non-adaptive behavior favors genes maintaining the migration pattern.

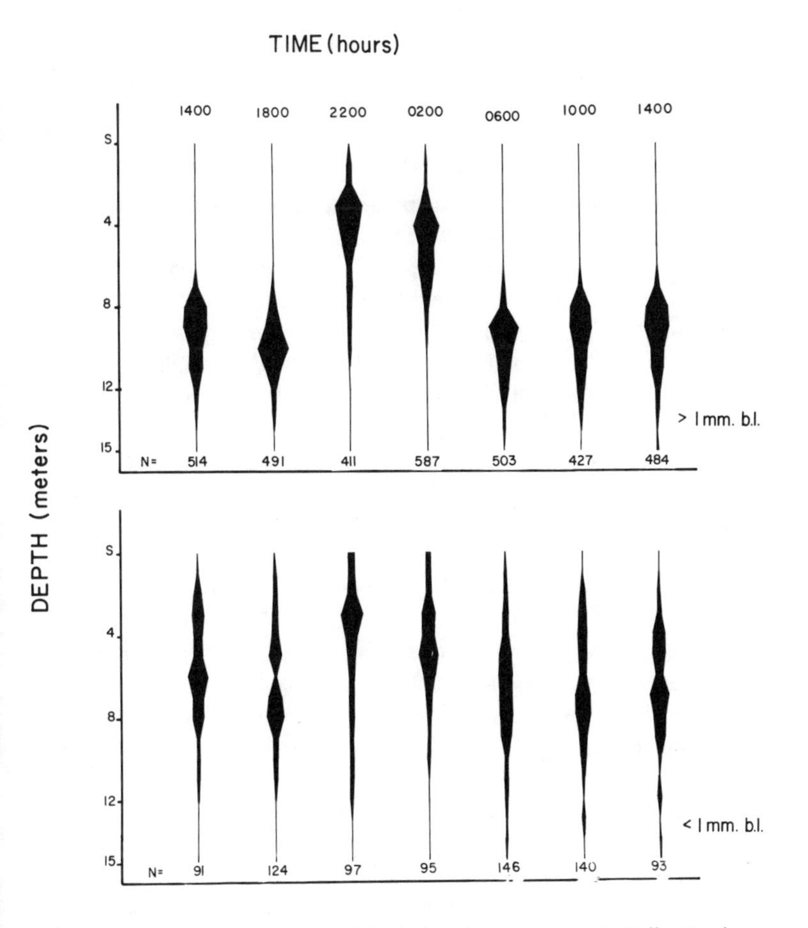

Fig. 25. Vertical migration pattern of *Daphnia galeata mendotae* in Fuller Pond, Connecticut. (From Suffern 1973)

Numerous examples of vertically migrating animals occur in the oceans (extensive reviews in McLaren 1963, Banse 1964). Pearre (1973) found that the arrowworm *Sagitta elegans* migrated to surface waters to feed at night and descended when sated (and see Conover 1968). Pearre concluded that the best explanation for the migration of these transparent animals was to avoid predation from visually hunting fish.

Predators such as fishes require light for more efficient prey capture.

The behavioral response of vertical migration can be an effective means of predator avoidance and is expected whenever visually dependent predators are present.

Visibility

The second cue for the gape-limited particle selector is prey visibility. In the presence of gape-limited predators, prey individuals with reduced amounts of visible body pigmentation will increase in proportion over evolutionary time. For instance, as has been demonstrated repeatedly, under intense predation pressures by planktivores the mean body-size of the prey population will decrease (Brooks and Dodson 1965; Galbraith 1967; Hrbáček et al. 1961; Hall, Cooper, and Werner 1970). A reduced overall body-size means a less visible carapace. Another response, occurring concurrently in cladocerans at least, is the reduction of the single most visible aspect of the animal, the large, black compound eye. Given a choice, planktivores will select larger-eyed cladocerans over their smaller-eyed, similar-sized sisters (Green 197l; Zaret 1972*a*). Consequently, the population mean size of the compound eye will decrease over time in the presence of planktivore pressure (Zaret and Kerfoot 1975).

Many species of freshwater planktonic animals are parthenogenic for a great part of the year. At the end of their seasonal cycle, females produce an overwintering sexual egg with a heavy carapace (ephippium) to resist unfavorable environmental conditions such as low temperatures, food shortages, or desiccation in ponds that dry out or freeze to the bottom during winter. These eggs, normally diploid, although they may develop without fertilization (Hutchinson 1967; Ratzlaff 1974), are considerably more heavily pigmented and more visible than the common haploid variety. Usually ephippia are produced in late summer in lakes, but the variability of this timing indicates that it is a flexible characteristic and amenable to evolutionary change. In general, the literature has related ephippial strategies to such environmental factors as temperature, reduced angle of light, or low food supply; and in the laboratories one can produce ephippia in normally parthenogenic populations by changing these environmental conditions. It is clear, however, that these environmental parameters are proximate factors rather than ultimate factors. That is, the environmental parameters act as reliable cues in indicating the appropriate time to produce ephippia; they are not the evolutionary reasons for producing ephippia at certain times of the year.

Planktologists have known for some time that when cladocerans such as *Daphnia* first start producing ephippia, examination of plankton-feeding fish will reveal an extremely high percentage of ephippial-bearing daphnids in their stomach contents. Some laboratory experiments have shown that fish will select ephippial *Daphnia* over non-ephippial adult *Daphnia* of similar body-size (Mellors 1975). Because daphnids are capable of producing ephippia at any time of the year, the timing of ephippia production may be an evolutionary response to fish cueing in on prey visibility (Zaret 1972*b*). In this way the gape-limited predators' dependence on visibility for feeding can affect aspects of prey morphology, pigmentation, and life history.

Motion

When a prey item is outside a fish's visible range, no amount of motion will make it visible. Once within the fish's visible range, however, motion can make one prey species item more conspicuous, and more likely to be eaten, than another species that is also within the visible field. Although it may be impossible for zooplankton species to completely reduce their swimming and feeding motions, some have evolved a motion that is distinctly less conspicuous than others'. For instance, the cladoceran *Diaphanosoma* and the calanoid copepod *Diaptomus* glide smoothly in the water with their long swimming antennae projected at right angles to the body and in the same plane; the cladoceran *Bosmina*, in contrast, appears to whir continuously through the water in tight circles. Workers such as Brooks (1968) have made similar observations concerning the comparative motion of various zooplankton groups.

A second type of zooplankton motion, the direct escape response, is not very effective for avoiding large gape-limited predators, but it is effective for avoiding size-dependent predators. Fish have considerable learning ability, and once they recognize the escape responses of even the fastest swimming copepods, they rapidly learn how to catch them. Confer and Blades (1975) found in laboratory studies that sunfish (*Lepomis gibbosus*) could remove copepods at up to 80 percent efficiency (successes per strike) after only a short period of acclimation. Hyatt (pers. comm.) found up to a 50 percent efficiency in kokanee (*Oncorhynchus nerka*, the landlocked sockeye salmon) feeding on *Diaptomus kenai*, a calanoid copepod that attains a body-size of approximately 3 mm. Only with very young and less mobile fish could a direct escape strategy be effective.

In Gatun Lake, Panama, the atherinid *Melaniris chagresi* rarely feeds on *Moina minutus*, a cladoceran with a maximum size 0.6 mm, and *Diaptomus gatunensis* with a maximum size of 1.3 mm (see fig. 12). In laboratory tanks with sufficient illumination *Melaniris* will immediately feed heavily on adult copepods put in with it, even though the copepod exhibits rapid escape responses. In contrast, the slower-moving *Moina* is never taken in the tanks, even though its size is larger than that of many preferred prey species. Only when black pigmentation is added to *Moina*, making them visible, are they immediately eaten by the fish (Zaret 1972*b*).

Escape responses may be successful if they enable the prey to move just outside the fish's visible field (see Drenner Strickler, and O'Brien 1978), but in general such responses are less effective than the ability to remain undetected. In addition, the energy costs associated with escape responses can be exceedingly bigh (Strickler 1977) and limit the prey to only a few escape attempts at any one time.

SIZE-DEPENDENT PREDATORS

Size-dependent predators, exclusively invertebrates, include such crustaceans as the cyclopoid copepods, especially the genera *Cyclops* and *Mesocyclops*; the calanoids, especially *Epischura* and *Diaptomus*; the cladocerans *Leptodora*, *Polyphemus*, and *Bythotrephes*; mysid and anastomid crustaceans; the dipteran larvae of *Chaoborus* and other insects, as well as rotifers, especially the genus *Asplanchna*.

LIGHT

Some size-dependent predators that may use light to capture prey, as suggested by their behavior or eye morphology, include *Polyphemus* (Brooks 1959), *Podon* (Bosch and Taylor 1973*a*, *b*), *Leptodora* (Cummins et al. 1969), and *Bythotrephes* (De Bernardi and Guissani 1975). When light-dependent predators of this type are affecting the prey populations, the prey might initiate migration responses to regions of low light levels.

Alternatively, vertical migration might be initiated by prey in response to nonvisual invertebrate predators, as suggested in the study by Fedorenko (1975*a*) on *Chaoborus trivittatus* and some of the crustacean prey of Eunice Lake, British Columbia. One of the larger calanoid copepods, *Diaptomus tyrelli*, can be eaten only by the large fourth instar of *Chaoborus*. These insect instars exhibit a vertical migration pattern in the lake, remaining near the bottom during daylight and migrating up at dusk to feed on *D. tyrelli* populations, which are located at that time in the lake epilimnion (surface) waters. Concurrent with the upward *Chaoborus* migration, the copepod undergoes a reverse migration downward. This behavioral response on the part of the copepod prey reduces the amount of time in which spatial overlap for *Diaptomus* and *Chaoborus* occurs. Because according to Fedorenko (1975*b*), the duration of spatial availability is one of the two factors that determine the diet of the *Chaoborus* predator, the reverse migration of the copepod may be an adaptive response to predation pressures.

Another example of prey migration in response to nonvisual, size-dependent predators can be seen in the studies by Narver (1970) on underyearling sockeye salmon in Babine Lake, British Columbia. Of the several crustacean species in the lake, three—the daphnid *D. longispina, Eubosmina coregoni*, and the predatory calanoid copepod *Heterocope septentrionalis*—are of greatest importance in the diet of the fish and are selected preferentially over other available prey items. In Babine Lake the sockeye feed most intensively at dusk and dawn, primarily on *D. longispina*, followed by *E. coregoni*, with *H. septentrionalis* which reaches sizes of up to 3 mm in body-length, lowest in numerical abundance in the fish stomach contents. These results are consistent with the diel vertical distribution of the fish and crustacean prey. The fish feed in the surface waters at dawn and dusk where the two cladocerans are located, whereas the copepod is found diurnally at 20 to 30 m depth. It is only at night that the predatory copepods undergo a typical migration upward, presumably to feed on *Bosmina*, although this relationship was not examined by Narver (1970). In contrast, *Bosmina*, associated with the epilimnion diurnally, undergoes reverse vertical migration in the evening, so that by night the bulk of the population is found at 12 to 20 m depths, even as their predators, the copepods, are vacating this region for the surface. As in the previous example, the coincidence of these predator and prey migration patterns probably relates to selective predation by size-dependent predators.

Hairston (1977) in his study of *Diaptomus* in Soap Lake, Washington, has also indicated the likelihood of invertebrate predation pressure favoring a reverse migration pattern for vulnerable prey. A contrary explanation was proposed by Bayly (1963), who found a correlation of low hydrogen ion concentration in Australian ponds where crustaceans exhibited reversed diurnal vertical migrations.

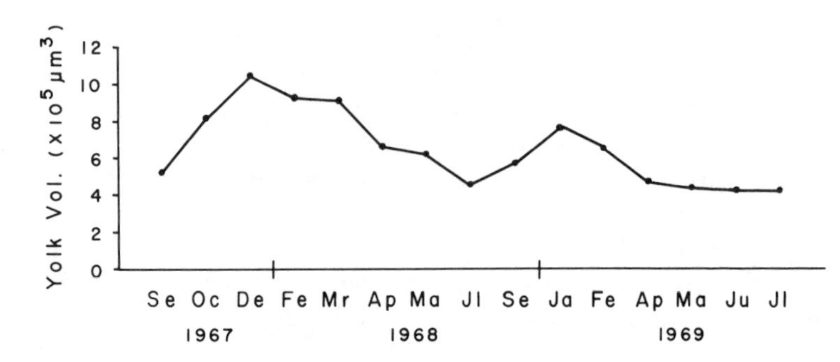

Fig. 26. Seasonal changes in amount of yolk per egg of *Bosmina longirostris* in Frains Lake, Michigan, 1967–69. (From Kerfoot 1974a, table 3)

Size

Prey individuals that attain body-sizes toward the right-hand tail of the size-dependent predator electivity curve (fig. 17) will experience increasingly reduced predation pressures. The growth rate of young can be increased by providing eggs with large amounts of yolk, thus enabling the young to reach adult size quickly. Kerfoot (1974*a*) studied populations of *Bosmina longirostris* under intense predation pressures by *Cyclops* and *Chaoborus* in Frains Lake, Michigan. He showed that prey females under heavy predation produced fewer eggs per clutch but provided each egg with a greater amount of yolk. This resulted in a larger neonate (newborn), which attained a larger size more rapidly. In the summer months of April through September, fish predation on *Bosmina* is significant, and during this time *Bosmina* eggs are smaller and with less yolk, presumably being less conspicuous for the visually dependent planktivores. In the months of October through February, however, fish predation is virtually absent. Size-dependent predators provide the important selective factors, and the yolk provided in each *Bosmina* egg is a significantly larger (fig. 26). Of all the lake cladocerans, only *Bosmina* has been observed to undergo this seasonal egg-size cycle, and it is unlikely that this is caused by physical conditions in the lake. Dodson (1974*b*) also noticed this seasonal egg cycle in *Daphnia middendorffiana* under intense invertebrate predation in alpine ponds.

Shape

Another strategy employed by a variety of prey species under size-dependent predator pressure is to increase in body-size, up to a certain limit, and produce specialized morphological structures that thwart the predator in its attempts to capture, handle, or swallow the prey. One well-known example is the cladoceran *Holopedium gibberum*, which attains a body-size of several millimeters and is encased in a transparent gelatinous sheath, appearing like a pea inside a transparent marshmallow (Fig. 27). Although this sheath is not a totally effective deterrent to fishes, which are capable of severely reducing *H. gibberum* populations (Stenson 1974; O'Brien 1975), it does prevent predation by invertebrates (Allan 1973).

One well-documented example of the altering of body-shape to reduce the effects of predation is *Bosmina longirostris* (Kerfoot 1975*b*, 1977*a*). By lengthening the antennules and mucrones (fig. 28), *Bosmina* losses from predation by the large calanoid copepod *Epischura nevadensis* in Lake Washington were reduced. After capturing a *Bosmina* individual, *Epischura* often manipulates the prey to expose the soft underparts, which it then scrapes out. During this manipulation the elongated antennules and mucrones of *Bosmina* may interfere with the predator's handling attempts. Furthermore, the mucrone or antennule the *Epischura* is holding on to may break off and allow the prey to escape, just as a lizard's tail may detach when it is attacked and held by an avian predator. For *Bosmina* this is evidently a very common means of successful escape during attack or handling because a large proportion of mature *Bosmina* females exhibit regenerated antennules and mucrones. During certain times of the year, 100 percent of the *Bosmina* population were found with regenerated structures (Kerfoot 1977*a*), indicating the high frequency of attack by *Epischura*. As a result, *Bosmina longirostris* populations living in areas with abundant *Epischura* have longer antennules and mucrones than those living in adjacent areas where the copepod is absent. It is assumed that these elongated structures may also increase the difficulty of initial capture.

Many populations of freshwater organisms that undergo cyclomorphosis exhibit such morphological changes as body-size, external protruberances, or other alterations of the exoskeleton, depending on the species (see figs. 9a, 9b, and 9c). The changes occur with each new generation rather than with individual change over one lifetime. It is almost always the limnetic representatives of not only the cladocera and rotifera but also the dinoflagellata that undergo cyclomorphosis over the year. Initial explanations that related these morphologies to flotation ability (see references in Dodson 1974*b*) were disproved (Jacobs 1967). More recent studies on cladocerans have suggested that these morphological changes are intimately related to predation pressures (Brooks 1965, 1968; Jacobs 1965,

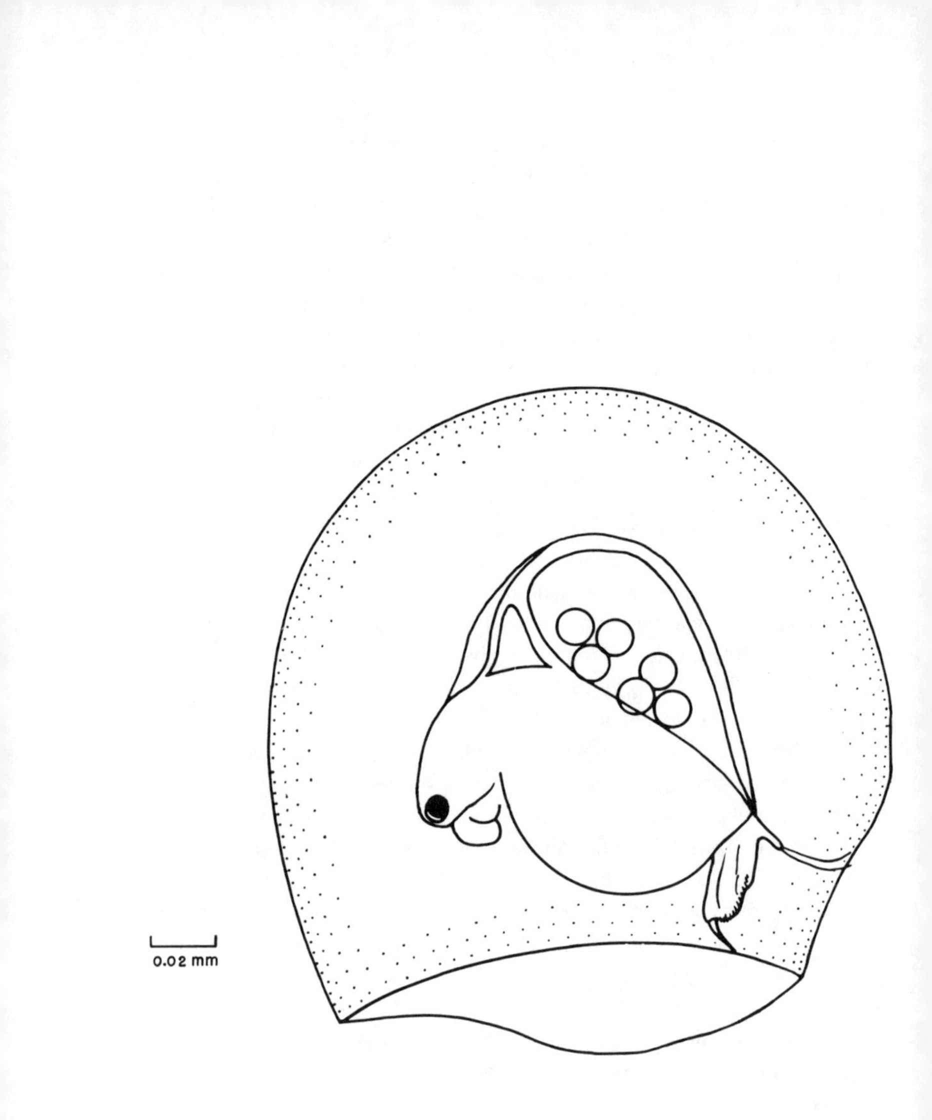

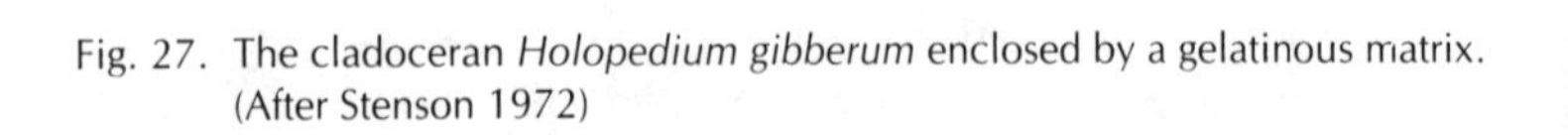

Fig. 27. The cladoceran *Holopedium gibberum* enclosed by a gelatinous matrix. (After Stenson 1972)

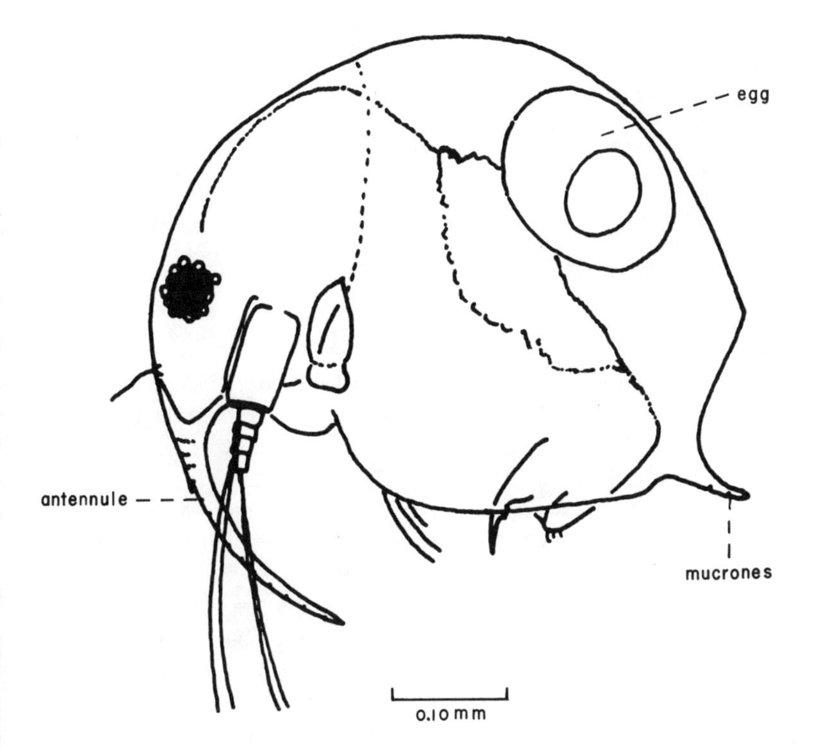

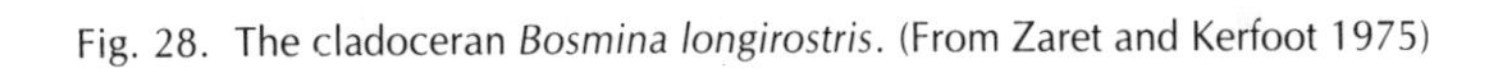
Fig. 28. The cladoceran *Bosmina longirostris*. (From Zaret and Kerfoot 1975)

TABLE 1. Adult-Body-size versus Helmet Development of North American *Daphnia*

Species	*Size (mm)*	*Helmet Development*
D. magna	3.0–4.0	None
D. similis	3.0–3.5	None
D. middendorfiana	2.0–3.0	None
D. pulex	2.0–2.5	None
D. rosea	1.6–2.0	None
D. catawba	1.6–2.0	Moderate
D. schodleri	1.2–2.0	Moderate
D. thorata	1.2–2.0	Moderate
D. dubia	1.0–1.5	Strong
D. galeata mendotae	1.0–1.8	Strong
D. laevis	1.0–2.0	Weak
D. longiremis	0.8–1.1	Strong
D. ambigua	0.7–1.0	Weak
D. parvula	0.8–1.0	Moderate
D. retrocurva	0.8–1.0	Strong

SOURCE: Brooks 1957.

1967; Halbach 1971; Gilbert 1967; Zaret 1972*a*, 1972*b*), and it has been suggested that in all of these groups the populations are responding with morphological accommodation to predation pressures (Zaret 1972*b*, 1975; Dodson 1974*b*).

One of the best-studied examples of cyclomorphosis is the genus *Daphnia* (Brooks' 1965). Many *Daphnia* populations develop what has been termed a "helmet" (see figs. 9b and 9c), an extension of the exoskeleton above the eye region. This helmet on the anterior end of the animal is correlated with an elongation of the posterior tailspine. Table 1 presents all fifteen species of North American *Daphnia* identified in Brooks's 1957 monograph on the family, arranged from the largest species to the smallest and indicating the extent of helmet development for each species. The body-size ranges represent mature instars only. Because this table is for the entire size range of species, there will be exceptions (e.g., *D. rosea* in Washington State does grow larger than 2.0 mm), but the values are useful for indicating trends.

The table shows clearly the correlation of helmet development with size. The five largest species, all but one of which mature at sizes of 2.0 mm and more, show no helmet development; those species that mature at sizes of l.2–2.0 mm show only moderate development when helmets are present in their populations; the four species that have strong helmet development mature at sizes in the range of 0.8 to 1.8 mm—all under the 2.0 mm size range. (Three species do not strictly fit this trend.) The value 2.0 mm is the approximate upper size limit above which these animals are relatively free from attacks by the largest size-dependent predator they commonly encounter, the phantom midge, *Chaoborus* (Dodson 1970, for *C. flavicans* and *C. nyblaei*; Fedorenko and Swift 1972, for *C. americanus* and *C. trivittatus*). Although *Chaoborus* can consume prey above 2.0 mm, its electivity drops off rapidly above this value. Predatory calanoid copepods are restricted to sizes below 2.0 mm (e.g., Dodson 1974*b*, for *Diaptomus shoshone*).

Information from Brooks's 1965 study of these animals, as well as from Green (1967), who measured helmeted and unhelmeted morphs of the same *Daphnia* species, indicates that the mean total length of populations with helmets and elongated tailspines can be increased up to 20 percent. This may be near the maximum upper limit, according to Brooks (although Burgis [1973] found that tail spines in daphnids of Lake George, Africa, attained up to 80 percent of the total length of an individual). The length of any individual, including helmet and tailspine, can be twice that of total body-length (i.e., without these two structures). The five species that mature at sizes of 2.0 mm and above will include mature individuals immune to predation by size-dependent predators and with no need for extensions via cyclomorphosis. When large, size-dependent predators are present, daphnids maturing in the 1.2 to 2.0 mm size range can persist only by the development of a moderate helmet and spine, which can bring their effective length above 2.0 mm. Finally, the smaller daphnids would need strong helmet development, thus increasing their body-lengths considerably, to be able to persist in the presence of these predators. Note that in the presence of smaller size-dependent predators, such as copepods, daphnids would not need to attain a size of 2.0 mm; this value is necessary to escape the largest invertebrate predators, such as *Chaoborus*. It is reasonable to suspect that the development of helmet and spine elongation correlates with daphnid body-size precisely because it is the smallest prey species that requires the greatest benefit from cyclomorphosis to reduce predation pressures.

Of the fifteen species in North America, three are apparent exceptions to this trend. *D. parvula* is small yet shows only moderate helmet development, perhaps indicating that it is not found in lakes with large invertebrate predators. The same may be said for *D. ambigua*, an uncommon

TABLE 2. Adult Body-size versus Helmet Development of European *Daphnia*

Species	*North American Analog*	*Helmet Development*
D. frigidolimnetica	*D. middendorffiana*	None
D. pulex	*D. pulex*	None
D. pulicaria	*D. pulicaria*	None
D. hyalina pelucida	*D. rosea*	None
D. hyalina hyalina	*D. rosea*	None
D. longispina		Weak
D. galeata	*D. galeata mendotae*	Strong
D. longiremis	*D. longiremis*	Strong
D. cristata		Strong
D. ambigua	*D. ambigua*	Weak
D. parvula	*D. parvula*	Moderate
D. cucullata	*D. retrocurva*	Strong

SOURCE: After Hrbáček

species that shows only weak helmet development and is usually restricted to hypolimnetic (bottom) lake regions (Tappa 1965). *D. laevis* matures at sizes of 2.0 mm; so its weak helmet development may not be an exception. It can also mature at smaller sizes; one needs to examine individual cases for clarification. One would predict a relationship between the size-dependent predator and the helmet size of daphnid species in different lakes.

This correlation also occurs in the European species of *Daphnia*. Table 2 lists twelve accepted species, arranged in decreasing size, together with Hrbáček's (pers. comm.) interpretations of the North American ecological analogs. The largest one, *D. frigidolimnetica*, corresponds to *D. middendorffiana*; *D. pulex*, a pond species; and *D. pulicaria*, a lake species never sympatric with *D. pulex*; and the two subspecies of *D. hyalina*, which correspond to *D. rosea*. These species never show helmet development. The species *D. longispina* approaches lengths of 3.0 mm and has weak helmet development. Species of the next group, including *D. galeata*, *D. longiremis*, and *D. cristata*, are all about the same size, and all exhibit helmet development in their populations. The third group, including *D. ambigua*, *D. cucullata* and *D. parvula*, shows weak to strong helmet development. Thus the same correlation of body-size and helmet development is seen, presumably for the same reasons.

The most spectacular examples of helmeted daphnids are individuals of *Daphnia cephalata* from Australia whose females attain sizes of up to 6 mm and possess grotesquely shaped carapaces. Differences among the perhaps ten species in the genus (Hebert 1977) in head and body-shape have been assumed to be solely environmentally determined (Bayly and Williams 1973). A reexamination of these populations to look for correlations between carapace shapes and pond predators would be benefical.

The rotifers also illustrate this effect of body-shape on predator success. Initial work showed that the herbivorous rotifer *Brachionus calyciflorus* exhibited a cyclomorphic change in the lengthening of its spines, especially the posterolateral pair, in the presence of the predator *Asplanchna brightwelli* (Beauchamp 1952*a*, 1952*b*). The elongation of spines was later shown to occur in other herbivorous species, including *Branchionus urceolaris* var. *sericus* and *Filinia mystacina* (Pourriot, 1964). The "*Asplanchna-substance*" was determined to be dietary, a-tocopherol, a vitamin-E compound (Gilbert and Waage 1967; Gilbert and Thompson 1968), which apparently is a good environmental indicator of future abundances of *Asplanchna* (Gilbert 1973*a*).

The adaptive significance of the elongation of the posterolateral spines in *Brachionus* was demonstrated by Gilbert (1967) in laboratory experiments with *Asplanchna girodi* and *A. sieboldi*. Gilbert designed feeding experiments in which he used adult *A. girodi* (mean size 0.75 mm), plus young (mean size 0.56 mm) and adult (mean size 0.87 mm) *A. sieboldi*. As prey for these he used young (0.15 mm) and adult (0.20 mm) *Brachionus calyciflorus*, separated according to whether they were spined or spineless. Gilbert compared the percentage of contacts leading to capture and the percentage of captures leading to ingestion by the *Asplanchna* predator. He found that *A. girodi* could capture 25 percent of the young *Brachionus* without spines but none of those with spines. In addition, *A. girodi* ingested only 33 percent of the spineless *Brachionus* it captured. Gilbert concluded that this *Asplanchna* species was rather inefficient at feeding on spineless *Brachionus* and was not at all capable of capturing or ingesting spined prey. The young of *A. sieboldi* were able to capture 89 percent of the contacted young unspined *Brachionus* but only 14 percent of the spined forms. Once its prey was captured, however, *A. sieboldi* young ingested 100 percent of both categories. For adult *A. sieboldi* feeding on *Brachionus* adults, 100 percent of the spineless *Brachionus* were captured and ingested but less than 75 percent of the spined animals. These studies conclusively supported the initial suggestions of Beauchamp

(1952*a*, 1952*b*) that the elongation of especially the posterolateral spines in *Branchionus* was adaptive as an anti-predatory device when *Asplanchna* were present. Gilbert's findings have been supported by later studies, including those by Halbach (1971) and Pourriot (1974), which found similar results with *A. brightwelli* and *Branchionus bidentata*.

In the actual encounter, *Brachionus* is able to move its posterolateral spines from the normal trailing position when swimming to one in which they are at right angles to its body. This occurs not from muscular control but from hydrodynamic pressure, which is increased when the animal withdraws its corona (Beauchamp 1952*a*), an event stimulated by contact with a predator. The result is a handling problem for *Asplanchna* and, as Gilbert (1967) noted, increased difficulty in grasping the prey. In addition, even though a captured prey will be ingested by *Asplanchna sieboldi* 100 percent of the time, sometimes as much as twenty minutes were required to get the captured *Brachionus* into the mouth because the spines were at right angles to the predator's mouth opening. This would greatly limit the feeding rate of *Asplanchus* on spined *Brachionus*.

The different morphs of *Asplanchna* itself demonstrate morphological changes. There are three basic forms: a small "saccate" type; an intermediate-sized humped, or "cruciform," type; and a large "campanulate" type similar to the saccate but attaining much larger sizes, up to 1.5 mm (fig. 29). Laboratory feeding studies by Gilbert (1973*b*) supported earlier suggestions that the humped forms were less likely to be eaten by cannibalistic campanulate forms (Hurlbert et al. 1972*a*). Gilbert showed that this was not strictly a function of size, because many of the humped animals that were not eaten were smaller than some of the nonhumped eaten saccates. It was the presence of the humps that decreased the rate of predator capture.

In the life history of *Asplanchna* the ephippial eggs give rise only to saccate females. These continue to produce parthenogenic females via amictic (diploid) eggs or females via mictic (haploid) eggs, which, if fertilized by males, produce ephippia, completing one aspect of the life cycle. However, if a-tocopherol is present in even minute amounts, the saccate individuals give rise to cruciform types; and if the level of a-tocopherol remains above threshold levels and additionally large-sized prey are abundant, which may or may not include conspecifics (Gilbert 1973*a*), gigantic campanulates appear.

The significance of these changes in the rather complex life cycle of

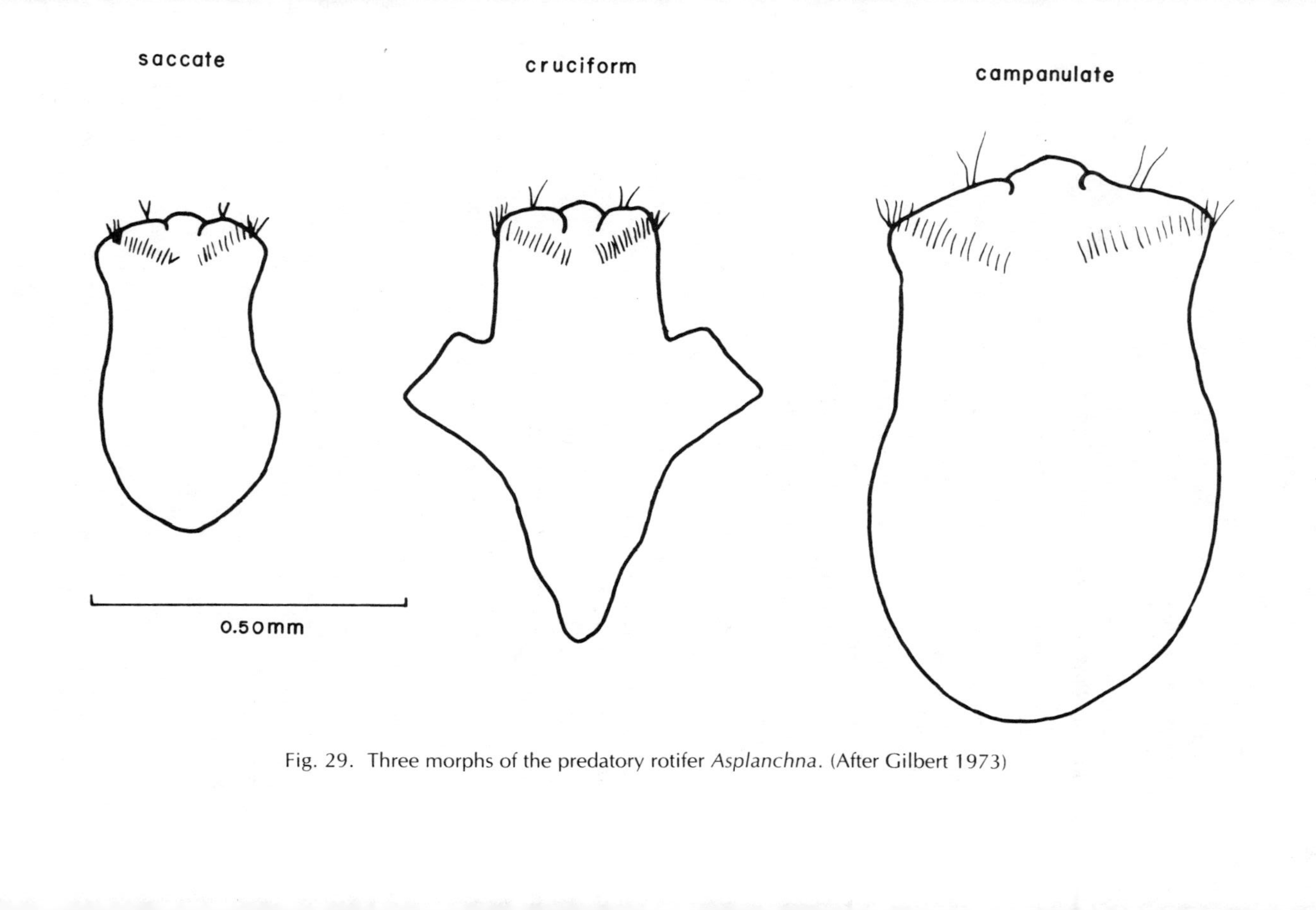

Fig. 29. Three morphs of the predatory rotifer *Asplanchna*. (After Gilbert 1973)

Asplanchna is that if prey are abundant, *Asplanchna*, given its size-dependent electivity curve, can consume the resources only if it attains a large enough size. Furthermore, predation by *Asplanchna*, on other rotifers will lead to an increase in the mean size of the remaining prey individuals, as Pourriot (1974) demonstrated for laboratory *Asplanchna* feeding on *Brachionus*. He showed that *Brachionus bidentata* populations co-evolving with *Asplanchna* had a larger mean size (exclusive of spines) than those grown in cultures without *Asplanchna*, presumably as a result of genetic changes resulting from the selection by *Asplanchna* of the smaller prey. Once large campanulates are present, however, they will, by random contact, encounter a certain number of conspecifics, including the males and saccate types involved in the sexual cycle. Thus the production of humps in the cruciform females and the haploid males is seen as a means for the population to take advantage of sudden abundances of larger-sized prey; and the humps will protect the production of the sexual eggs, which may be the sole source of the following year's *Asplanchna* population. This is especially important because the campanulates never produce mictic females and have no sexual cycle. In this case the levels of a-tocopherol, a compound present in most algae, which is necessary to induce the changes from saccate to cruciform to campanulate, may be viewed as the best indicator of future algal abundances and hence the future abundance of herbivorous prey. Thus, as algal resources increase, herbivores (such as *Brachionus*) also increase. As increasing numbers of prey are ingested and the level of a-tocopherol present in the *Asplanchna* rises, the predator responds to the increase in herbivore populations in ways adaptive for the exploitation of prey.

Motion

Prey motion is the fourth cue that size-dependent predators may utilize. In the presence of fishes the best strategy for prey species is to be inconspicuous; in contrast, visibility is important for only a minority of size-dependent invertebrates. Because copepods are dependent on hydrodynamic information for prey capture, as presumably are others, including *Chaoborus*, any prey able to negate these sensing abilities will have a successful means of escape. If a copepod attacks and misses or is not able to maintain contact with the small cladoceran *Bosmina longirostris*, the escaping *Bosmina* will respond by tucking its swimming appendage in behind its protective antennules. Once the animal ceases swimming, it drops passively, sinking down and out of the range of the copepod, which will itself exhibit a series of looping somersaults in attempts to relocate the prey item by bumping into it (Kerfoot 1977*a*). This cessation of activities by the prey, halting the transmission of hydrodynamic information, seems to be an effective strategy for prey that is very small relative to the size of its invertebrate predators (R. E. Zaret, pers. comm.).

Direct escape responses of various common zooplankton species—some prey swim away from the predator—were studied in the laboratory by Szlauer (1965). Using glass tubes that he lowered into a beaker containing zooplankton species, Szlauer recorded escape ability by determining the percentage of a given species that effectively escaped before the tube could be lowered over it. Those species with the highest ability to escape were the adult copepods *Eudiaptomus graciloides* and *Thermocyclops oithonoides*, which showed escape percentages of 89.4 (males), and 83.8 (females) and 78.7 (males), respectively. Next came the copepodid stages, ranging from 70.8 to 56.5 percent, and the cladoceran *Diaphanosoma brachyurum* at 70.8. Following these were *Daphnia cucullata*, 47.9 (females) to 35.7 (males); *Eubosmina coregoni kessleri*, 34.5; *Chydorus spaericus*, 24.0; and the rotifers *Polyarthra* spp., 21.7, and *Notholca longispina*, 17.7. The copepod nauplii (11.9), *Keratella cochleris*, (9.5), and *Keratella quadrata* (1.5) exhibited poor escape abilities. Finally, a 0.9 percent escape ability was shown by the rotifer *Synchaeta* spp. and, as discussed, zero percent by the cladoceran *Bosmina longirostris*. These experiments demonstrate the wide range of direct escape abilities exhibited by a number of zooplankters, and there appears to be a direct correlation of escape ability with prey size. Experiments with artificial "siphon-tube" predators performed by Singarajah (1969, 1975) for various marine larvae led to similar conclusions.

The energy expenditure necessary for escape movements may be considerably more than first thought. Vlymen (1970) calculated that the energy of normal swimming movements for the marine copepod *Labidocera trispinosa* was small compared to its basal metabolic rate. For high acceleration hops, however, as used by copepods in escape responses, the cost is evidently large (Strickler 1977). Thus, direct escape movements appear to be a high-cost strategy for escaping predators, greatly limiting their frequency. Further, the direct escape response is more effective against smaller invertebrate predators than against the much larger and more agile fish, which can learn readily to take even the largest, fastest swimming

TABLE 3. Prey Cues Received by Freshwater Predators and Evolutionary Responses to Predation Pressures

Predator	Dependent On	Prey Response
Gape-limited predator		
Fish (*Melaniris*)	Light	Behavioral (vertical migration)
Salamander (*Ambystoma*)	Visibility	Reduced pigmentation (i.e., body-size, eye-size reduction)
Insecta (*Notonecta*)	Motion	Reduced motion
Size-dependent predator		
Cladocera (*Leptodora, Polyphemus*)	Light	Behavioral (vertical migration)
Calanoid copepoda (*Epischura*)	Size	Increased body-size
Cyclopoid Copepoda (*Cyclops*)	Shape	Morphological adaptation
Diptera (*Chaoborus*)	Motion	Escape responses

copepods. For these reasons, I think it is reasonable to conclude that the direct escape responses of these animals have evolved for size-dependent predation pressures. Table 3 summarizes prey responses to the two types of predators.

6. Community Models

In any system some components have more influence or predictive power than others. By understanding the structure and function of key components, we can understand the organization of the entire system (Holling 1961). If we can identify the dominant predators in a lake and the type and direction of their selection, we can predict the prey species that should be present. When gape-limited predators dominate, prey will be characterized by (1) vertical migration behavior; (2) small body-sizes, small compound eyes, and other morphological attributes that result in reduced visibility; and (3) an inconspicuous means of locomotion. The predominance of size-dependent predators will favor the presence of prey characterized by (1) vertical migration behavior; (2) life histories favoring rapid initial growth; (3) large adult body-sizes, (4) morphological adaptations that present capture or handling problems or both; and (5) direct escape movements. When species morphology is used as a basis for taxonomic identifications, this kind of information can be used to develop predictions of community species composition.

Three additional assumptions relate to the form of interaction between the two categories of freshwater predators. *First, gape-limited predators regulate the densities of the size-dependent predators through predation.* The interplay between visibility and motion among the various invertebrate predators present means that some size-dependent predators are more likely to occur in the presence of piscine planktivores than others. Fish will immediately consume and remove any of the large opossum shrimp, *Branchinecta*, or mysids such as *Neomysis*, so that only rarely (e.g., in some very large lakes) will these invertebrate predators be found sympatrically with piscine planktivore populations. Other predatory cladocerans, such as *Leptodora* or *Bythotrephes*, because of their relative transparency, can occur simultaneously with fish in spite of their attaining large sizes. When cladocerans such as *D. pulex* are also present, fish may select them over *Leptodora* and *Bythotrephes* because the daphnids are relatively more conspicuous. Only when these largest *Daphnia* are removed and replaced by smaller daphnid species do fish turn their full attention to the predatory cladocerans (Fuchs 1967; Cummins et al. 1969; De Bernardi and Giussani 1975). In contrast, there is usually a strong negative correlation between fish and *Chaoborus* populations (Pope, Carter, and Power 1973; von Ende 1979). The partially transparent *Chaoborus* is often preyed upon by fishes in preference to cladocerans (Allan 1973), probably because it has a very conspicuous darting motion, which may serve as the basis for a searching image. The regulation of invertebrate predators by gape-limited predators is not determined simply by body-size consideration.

Second, the interplay between visibility and motion occurs for fish feeding on cladocerans and predatory copepods. *Given a choice of large daphnids or larger-sized copepods (up to a body-size ratio of 1 : 1.5), fish will selectively remove the cladocerans.* As mentioned previously, in Brooks's 1968 laboratory experiments the alewife *Alosa pseudoharengus* removed the smaller *Daphnia catawba* before the larger *Epischura nordenskioldi*. In lakes, when large daphnids are not available, equally large copepods are ignored and many fishes switch instead to benthic prey (Brooks 1968). The preference of planktivorous fishes for cladocerans over large copepods has also been shown in several studies (Ivlev 1961; Archibald 1975). Losos and Hetesa (1973) indicate that carp fry grow better when feeding on daphnids than on copepods.

A third important relation among the predators is the prey preference of size-dependent invertebrates. Such limnetic copepod predators as *Epischura* and *Diaptomus* have probably evolved to exploit the relatively slow-moving, easily captured limnetic cladocerans. Given a choice, these size-dependent predators will remove more cladocerans than copepods of a similar size. With *Chaoborus*, the reverse is true because it normally swallows its prey whole (Pastorok 1978). The slender copepods can be ingested more rapidly and more easily than egg-shaped cladocerans, and *Chaoborus* prefers the former as prey (Swift and Fedorenko 1975). In addition, it appears that high *Chaoborus* densities in lakes are correlated with low levels of the other size-dependent predators, *Polyphemus*, *Leptodora*, and

Epischura (Pope and Carter 1975), at least in part because of predation by *Chaoborus* on the crustaceans.

If we know the level of predation intensity of the gape-limited predator we can predict the coexisting prey species and level of predation intensity of the size-dependent predator. By considering the interactive predators alone and together we can make predictions about the resulting prey type.

To construct the model, let us consider gape-limited predation pressure as a series of states increasing in intensity and ranging from a zero state (absence of gape-limited predators) to +6 (extremely intense predation). An increase in predation intensity may correspond to: (1) an increase in the number of planktivore individuals in the lake; (2) an increase in the total number of planktivore species present; (3) a change to more efficient lake planktivores; (4) a change in planktivore population size distribution as the young of the year appear; (5) a switch by predators to plankton from other food resources; or (6) a change in physical conditions in the lake, such as seasonal alterations that may allow increased planktivore efficiency.

A decrease in predation intensity may correspond to: (1) a decrease in the number of planktivore individuals in the lake; (2) a change in one planktivore population as the young of the year mature and switch to non-plankton food; (3) migration of planktivores from the lake system; or (4) a change in physical conditions in the lake, decreasing planktivore efficiency. In order to simplify the initial description, the first model considers copepods as the size-dependent predator (SDP) and only cladocerans as the prey. The species pool comes from the area extending from New England to the Great Lakes. This first application, Sub-model I, is presented in table 4 and discussed below.

SUB-MODEL I

State 0

In State 0 there is a complete absence of gape-limited predation, neither fishes, salamanders, nor insects, so that copepods will emerge as the dominant predator. Copepod predators have an electivity curve that favors prey with rapid initial development and large adult body-size, so that prey adults are outside the maximum size-range copepods are capable of eating. Once the cladoceran prey population is established, with mature females above the critical size threshold, some cladoceran individuals will achieve maturity to reproduce. Because prey density is critical in limiting

TABLE 4. Sub-model I. Northeastern United States

Gape-Limited Predator	*Cladoceran Prey*		*Size-dependent Predator*
Fish	*Cladoceran*	*Size (mm)*	*Copepod*
0	*Daphnia pulex*	2.0–2.5	+2
+1	*D. catawba*	1.6–2.0	+3
+2	*D. galeata mendotae*	1.0–1.8	+4
	D. longiremis	0.8–1.1	
	D. retrocurva	0.8–1.0	
+3	*D. ambigua*	0.7–1.0	+3
+4	*Ceriodaphnia lacustris*	0.4–0.7	+2
+5	*Eubosmina corogoni*	0.4–0.6	+1
+6	*Bosmina longirostris*	0.3–0.5	0

size-dependent-predator populations, the predatory copepods cannot become very abundant until the prey population attains a sufficiently high density. The predation intensity of these large, size-dependent predators will be low, moving from States +1 to +2 as prey numbers exhibit seasonal changes.

Given the dominance of copepod predation, we would predict the cladoceran prey to be the largest local species of cladoceran, such as *Daphnia magna*, which attains sizes of up to 3 mm (see table 1). This species and its ecological analog, *D. similis*, with which it is never sympatric, are quite different from the other members of the genus, a fact recognized by taxonomists who place them in a separate subgenus, *Eudaphnia* (Brooks 1959). In addition, *D. magna* and *D. similis* are characteristic not of lakes but of such temporary bodies of water as transitory ponds.* Both species attain a very large size, far above the 2 mm threshold necessary to make them less vulnerable to predation by the SDPs normally associated with lakes.

If one assumes *D. magna* to be the most efficient competitor among herbivorous zooplankton (e.g., Brooks and Dodson 1965; Burns 1969*b*), it is unclear why this animal is not found in lakes without fish. It appears

* The apparent habitat exception for *D. magna* is the London, England, reservoirs where it is the dominant crustacean species numerically and in biomass. These basins, however, are relatively shallow and heavily influenced by the river Thames, whose waters are stored in the reservoirs (Cremer and Duncan 1969). The constantly high detrital levels and absence of GLPs (Duncan 1975) make these reservoirs ecologically similar to the pond habitat where *D. magna* is normally found.

that *D. magna* (and presumably *D. similis*) has evolved three characteristics that make it especially well adapted for temporary ponds. First, its rapid filtering ability enables it to take advantage of high algae densities, occurring in dense blooms, to accumulate energy and grow rapidly to a large size. Second, it is adapted to feeding on detritus or other bottom-associated food sources characteristic of shallow ponds. Third, these eudaphnids probably attain such a large size not because of lake predators but because of those predators associated with temporary ponds, such as larvae of odonates, coleopterans, or other predatory insect groups.

Excluding from consideration the two largest North American daphnid species because of their specialization for pond life, we predict that the next largest species, *D. pulex* (2.0–2.5 mm) or *D. middendoriffiana* (2.0–3.0 mm), will be dominant in lakes without GLPs and with predatory copepods. Thus, with gape-limited predation at State zero, we expect size-dependent predation at State +1 or +2 and the cladoceran prey to be *D. pulex*.

State + 1 to + 2

The +1 state can be represented by a salamander, by a fish that is a generalized planktivore, or by a moderately efficient planktivore having a low population abundance. Given a lake dominated by a large predatory copepod and a large cladoceran, such as *D. pulex*, the planktivore will select the cladoceran over the copepod, so that initially predation will fall most heavily on *D. pulex*. As the planktivore pressure increases, the behavioral response of vertical migration occurs as a rapid means of adjustment. If such lake conditions as morphometry and water transparency allow, this behavioral accommodation can reduce the feeding efficiency of the planktivore sufficiently to allow persistence of the large cladocerans.

The relationship of vertical migration patterns in the same lake with increasing GLP intensity is illustrated in Fuller Pond, Connecticut, a small kettle lake studied by John Brooks and his associates from Yale University. When this lake was first sampled by Brooks in 1941 (unpublished data), he found populations of two daphnid species, (1) *D. pulex* adults with mean sizes 2.0–2.5 mm and the smaller *D. galeata mendotae* with a mean adult size 1.0–1.8 mm. At this initial date, the ratio of *D. pulex* to *D. galeata mendotae* was 1.5 : 1. In addition, Brooks found that more than 80 percent of the *D. pulex* were at a depth of at least 10 m, with no "large individuals" (meaning presumably adults) at a depth of less than 8 m. In contrast, more than 70 percent of the *D. galeata mendotae* were at a depth of 5 m, with no "large individuals" below 8 m, that is, almost all adults were at that same 5 m depth. The fishes in the lake included the long-eared sunfish (*Lepomis auratus*), the yellow perch (*Perca flavescens*), the pumpkinseed (*Lepomis gibbosus*), and possibly the smallmouth bass (*Micropterus dolomieu*), all of which were introduced by 1906.

The golden shiner, *Notemigonus crysoleucas*, probably was introduced to Fuller Pond in 1952, and it is evidently this species, a more efficient planktivore than any of the centrarchids, that caused significant change in the lake (Suffern 1973). Preliminary studies undertaken in 1968 by Burns (1968*b*) and subsequently continued by Suffern (1973) confirmed a major change in the composition of the limnetic zooplankton. Populations of *D. pulex* were no longer present; *D. galeata mendotae* was now the dominant cladoceran and had acquired a striking vertical migration pattern not present in 1941. The disappearance of *D. pulex* was not related to changes in water conditions. Burns (1968*b*) demonstrated by lake enclosure experiments that *D. pulex* could persist in Fuller Pond in situ for at least three weeks if contained within nylon netting cages. In addition, Burns concluded that the daytime depth of *D. galeata mendotae* was not causally determined by physical factors such as temperature, light, oxygen concentration, or turbulence. The minimum carapace length of a mature instar of *D. galeata mendotae* decreased during the sampling program from 0.87 mm on June 7 to 0.63 mm on November 17, indicating intense predation by the gape-limited predator, the golden shiner, which is now the dominant planktivore in the lake (Suffern 1973).

The conclusion is that in 1941 *D. pulex* persisted through vertical migration behavior, and the smaller *D. galeata mendotae* was relatively free from predation by planktivores. When the golden shiner was introduced and predation intensity increased, *D. pulex* was exterminated. Following this, the predation pressures on *D. galeata mendotae* greatly increased, and only those individuals that exhibited migration patterns survived. This illustrates how vertical migration might be favored by increasing gape-limited predation, and how the effectiveness of this behavior pattern is determined by the interaction of predation intensity, prey size, and the lake physical environment.

State +1 is defined as a level of GLP intensity that eliminates large cladocerans from the environment. That is, as fish predation increases, the prey individuals can adjust initially with the behavioral adaptation of vertical migration. If fish predation is sufficiently intense, the largest, most visible individuals will be removed, and only individuals than can mature at smaller sizes will persist. This decrease in mean size of fertile instars will continue with increased predation pressures until fish are able to crop the *D. pulex* population to the extent that no mature instars survive, meaning that the population will become exterminated. This transition has been shown in lake studies for both *Daphnia* (Galbraith 1967) and *Bosmina* (Stenson 1976) under fish predation. The niche for a large cladoceran will no longer be available and will be replaced by a niche for a smaller form, which will allow another species to enter the system.

If the planktivore is relatively inefficient, so that it attains a predation intensity of only +1, *D. pulex* can be replaced by a prey of slightly smaller total visibility such as *D. catawba* (mean mature size 1.6–2.0 mm); the predator's net energy obtained per unit time is thus lowered, and the system can come into balance at this new state. If the planktivore attains a +2 state, this prey niche is no longer available, and the species of daphnid expected is from a still smaller group, *D. galeata mendotae, D. longiremis*, or *D. retrocurva*, all of which mature at sizes between 0.8–1.8 mm. The particular species will depend on predation intensity; as Hall (1964) found, when fish reduced the population of *D. galeata mendotae* in Base Line Lake, Michigan, this species was subsequently replaced by the somewhat smaller *D. retrocurva*. These three daphnids are lumped together at State +2 because they are within a range of 1.0 mm and are similar in their life-history patterns.

Prey density is especially important for SDPs because of their dependence on the random encounter of prey for capture. With the establishment of a daphnid species whose mature individuals have a smaller mature instar body-size, there will be an increase in the number of individuals in the smaller-size categories. This shift in daphnid size distribution will favor populations of SDPs, leading to an increase in the intensity of predation corresponding to States +3 and +4 for SDPs.

The direct correlation of increasing fish predation and an increase in numbers of smaller-sized prey is shown by Galbraith's 1967 study, which examined the zooplankton and fish populations of several Michigan lakes. In Stager Lake, the control lake, the fish species composition remained basically the same over the study period. In Sporley Lake predation intensity increased by the addition of several planktivores over the same time period. Galbraith was concerned with changes in the several daphnid species in Sporley Lake and considered two sizes—those greater than 1.3 mm

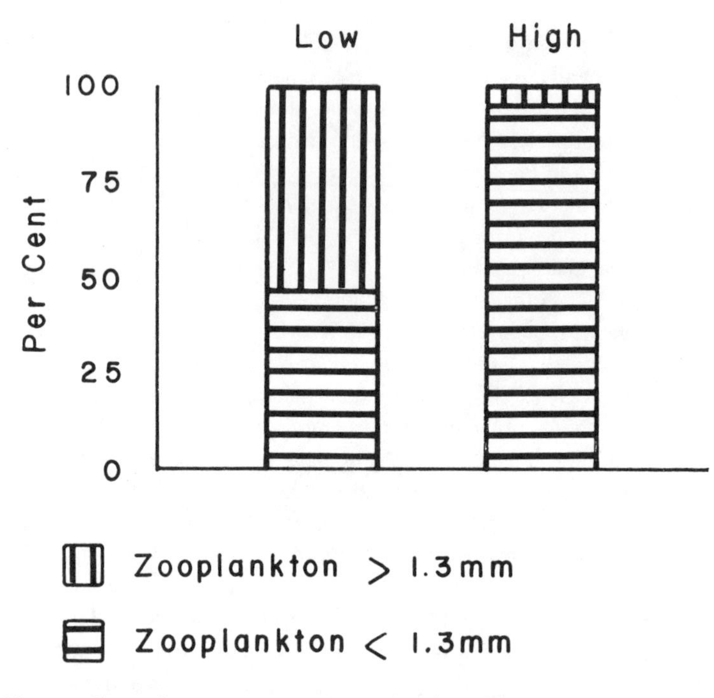

Fig. 30. Change in percent composition in Sporley Lake, Michigan, of two size fractions of genus *Daphnia*, comparing time of low planktivore predation levels (1958–60) with high levels (1964–65). Increase in total *Daphnia* densities also associated with latter time period.

and those less than 1.3 mm length. Over this time period a dramatic increase in the density of animals per liter occurred, the values for 1964–65 reaching up to four times those found for the period 1958–60. Furthermore, even though the total daphnid density increased, the total volume of *Daphnia* per liter actually decreased over the same period as individuals of larger species were replaced by smaller ones. Using Galbraith's data, the average summer proportion of daphnids less than 1.3 mm increased from 46.2 percent of the population in 1958–60 to 95.3–100 percent for some dates in 1964–65 when the study ended (fig. 30).

These results provide one explanation for the observation that predatory cladocerans are found only when fish are present (e.g., Pope and Carter 1975). Not only is the cladoceran prey species of a size that falls within their capture range, but more important, the density of small-sized prey has increased manyfold. In Galbraith's 1967 study, for instance, the density of small daphnids with GLP reached eight times that found during the years when fish were absent. Not surprisingly, the population of predatory copepods was also most abundant in plankton samples during these years (Kerfoot, pers. comm.). *L. kindti* first appeared in Sporley Lake in the later years (Galbraith, pers. comm., in Dodson 1970). We predict the appearance of other size-dependent predators, such as the cladocerans *Leptodora* or *Bythotrephes*, as the GLP reaches State +2.

Returning to the model, State +2 signifies moderate planktivore predation, primarily on the cladocerans. The population and predation intensity of predatory copepods are at their zenith because fish predation is light and the number of small-sized prey is high. Both GLP and SDP forces are present, and prey, consequently, should have characteristics adaptive for both types. The cladocerans of the 0.8–l.8 mm size range, *D. galeata mendotae, D. longiremis*, and *D. retrocurva*, have a reduced visible body-size in response to the visually dependent gape-limited predators (see Brooks 1968); they exhibit also a great reduction in the size of their compound eye—up to 80 percent relative to nonhelmeted morphs of the same species (Zaret 1972*b*). In addition, the presence of a helmet and an elongated tailspine in these daphnid species enables thier total carapace to exceed the handling capabilities of invertebrate predators.

State + 3 to + 4

The feeding efficiency and removal rates of salamanders (or insects) probably never attain the level of State +3 to +4 predation intensity, so that from this point only planktivorous fish are considered. As the level of GLP intensity increases toward +3, the initial effect will be on the largest crustaceans. Vertical migration by these species could be expected and possibly the presence of a smaller daphnid species such as *D. ambigua* (mean mature body-size 0.7–1.0 mm). As the mean size of available cladoceran prey drops below 1.0 mm, these fish may switch to the SDP, because many fish appear to rapidly lose interest in cladocerans as they decrease toward body-sizes of 1.0 mm (Brooks 1968, 1969). As fish predation increases, vertical migration patterns may occur among the invertebrate predators, as well as reverse patterns for their prey. This results in rather complex interactions among the various crustaceans of the community, now migrating in response to at least two different predation pressures. As fish begin selecting the SDPs as prey, a gradual decrease in the invertebrate predator populations should occur and consequently, the predation pressure exerted by them, returning from State +4 to State +3.

At State +4 the predation intensity by fish has reached the level where no daphnid population can survive (i.e., maintain mature instars). The copepod populations are being diminished also, and their predation intensity is dropping farther to +2. (In general, copepods appear able to migrate greater distances and more quickly—and thus more effectively—in their vertical patterns than most cladocerans.) To survive under this intense fish predation, prey even smaller than *Daphnia* are expected. The next smallest cladocerans, the genus *Ceriodaphnia*, which includes several species, are 0.4–0.7 mm maximum size. These cladocerans, especially the immatures, are fed upon by copepods and other size-dependent predators, but because the intensity of invertebrate predation is now greatly reduced, the smaller *Ceriodaphnia* species can persist. Morphological adaptations in these prey species may occur. Some species in this size range, including the common *Ceriodaphnia lacustris* or the tropical *C. cornuta*, possess fornix spines and sometimes head spines, which may be considered anti-predator devices (see fig. 9a). As the largest SDPs, such as *Epischura*, are removed by fish, they will be replaced by such smaller predators as *Diaptomus* as a response to GLP. These smaller predatory copepods will have a reduced prey-size range that they can effectively capture and handle, which also favors the persistence of these small-sized cladocerans in the system.

State + 5 to + 6

In State +5, fish predation is very intense. This may correspond to extremely efficient planktivores or to a combination of several fish species. The fish are capable of feeding on *Ceriodaphnia* as small as 0.3 mm, and initially vertical migration may occur in the cladoceran populations. Because copepods are also being consumed by fishes, their level of predation intensity is lowered to a +1 state. Under these conditions of near to complete absence of predatory copepods, very small cladocerans should dominate, such as the smallest, limnetic cladocerans, the Bosminidae. Often considered littoral inhabitants, Bosminidae populations may expand to open-water lake regions when GLP is intense enough to remove other limnetic crustaceans (Brooks and Dodson 1965; Burbidge 1974; Noble 1975). When copepod predation is present, anti-copepod morphological appurtenances may be seen in the Bosminidae, as has been demonstrated for *Bosmina longirostris* (Kerfoot 1977a). Another genus, *Eubosmina* (see Deevey and Deevey 1971), responds with a much more strongly developed antennule and tailspine, features that are especially prominent in young *Eubosmina* individuals upon whom copepod predation would be most intense. For example, the ratio of antennule-length to body-length decreases with age in *Eubosmina tubicens* in Gatun Lake (Zaret unpublished data). Because *Eubosmina* is the larger of the two genera and appears better equipped to handle SDP pressures, it might appear at State +5. As GLP increases toward State +6, *Bosmina* should appear with elongated antennules and mucrones; and when fish predation is so intense that size-dependent predators are eliminated, *Bosmina* appears without antipredatory devices.

This is the last defined state, thus completing Sub-model I. Several conclusions can be made at this point concerning the actual structure of the model.

First, although the introduction of a second planktivore or the hatching of young fish is a distinct act, a longer time is required for a zooplankton community dominated by *D. pulex*, or one with *D. pulex* undergoing vertical migration patterns and *D. galeata mendotae* also present, to become a lake where *D. pulex* is absent and *D. galeata mendotae* dominates. We may conveniently categorize the different communities captured in the zooplankton sample during this change as "multiple-equilibria" (Sutherland 1974), but this is a verbal contrivance for what is a much more dynamic process. The previous model is being used to illustrate the very well-defined trends, their sources, their mechanisms, and the interactions that result in the communities, but the assignment of six static states is really for scientific convenience.

Second, a real community may vacillate between two or perhaps even three states over time or space, that is, within the same lake. In Union Bay, Lake Washington, heavy fish predation near the margins of the lake has resulted in a community of State +5 to +6, whereas in the waters farther from shore the fish are absent and a State +3 or +4 situation exists (Kerfoot 1977a). Alternatively, we might have a GLP that is restricted to surface waters, resulting in a surface community of State +6 and lower in the water column a community at State +2. In general, however, in a large lake it is difficult to maintain any distinct separation of states because of constant water mixing; one would never expect to see *Daphnia pulex* and *Bosmina longirostris* in one lake at the same time unless the lake has been undergoing recent radical changes. Seasonal changes in the lake may also

TABLE 5. Sub-model II

Gape-Limited Predator	Predatory Copepods		Herbivorous Copepods		Predatory Rotifers
	Large	Small	Large	Small	
0	+2	0	+4	0	0
+1	+3	0	+3	0	0
+2	+4	+1	+3	0	0
+3	+3	+2	+2	+2	0
+4	+1	+3	+1	+3	0
+5	0	+2	0	+2	+1
+6	0	+1	0	+1	+3

move the system quite rapidly from one state toward another, whether as a result of physical changes or the hatching of fish larvae, which then enter the limnetic water to feed.

SUB-MODEL II

Sub-model II is a modest extension that attempts to integrate the herbivorous copepods within a sub-model in order to describe their interrelationships. As fish predation exterminates such large, predatory copepods as *Epischura*, there are cascading effects on the herbivorous and smaller predatory copepods. The effective escape responses of large herbivorous copepods prevent them from being taken readily by predatory copepods or other invertebrate predators. When these large SDPs dominate, however, smaller herbivorous copepod species, which are most vulnerable to predation, should not occur.

With increasing fish predation, the large predatory copepods are removed, and future fish predation will fall on the large herbivorous copepods, such as *Diaptomus* or the cyclopoid *Eucyclops*. These will then be replaced by smaller-sized herbivorous copepods, even as the large predatory copepods are replaced by smaller predatory ones (or those normally restricted to the littoral region, such as cyclopoids). Again, the smaller herbivorous copepods occur only when large predatory copepods are not present. Sub-model II incorporates this logical extension from the original model (table 5).

Sub-model II also includes the rotifers. The ecological role of rotifers in lake systems is still poorly understood. We have only the pat generalization that their abundances are inversely correlated with those for clado-

TABLE 6. Sub-model III. Northwestern United States

Gape-Limited Predator	Cladoceran Prey		Size-Dependent Predators	
Fish	Cladoceran	Size (mm)	Copepod	Chaoborus
0	*Daphnia middendorffiana*	2.0–2.5	0	+5
+1	*D. rosea*	1.6–2.0	+1	+3
+2		1.2–2.0	+2	+2
+3	*D. galeata mendotae*	1.0–1.8	+3	+1
+4	*Ceriodaphnia* sp.	0.4–0.7	+1	0
+6	*Bosmina longirostris*	0.3–0.5	0	0

cerans, with whom they are probably competing for food (see Hurlburt et al. 1972*b*). One well-studied rotifer genus, *Asplanchna*, which can attain sizes of up to several millimeters, feeds on cladocerans when rotifers are scarce (Hurlbert et al 1972*a*). When present with intense fish predation, *Asplanchna* populations exhibit high transparency as an adaptation to visual predation (e.g., Nilsson and Pejler 1973). *Asplanchna* feeds mainly on very small cladocerans and is an important crustacean predator during times of heavy fish predation, when the density of small cladocerans is high. When the crustacean fraction is low, coinciding with a high density of total rotifer populations, *Asplanchna* switches to feeding on rotifers. *Asplanchna* is added to Sub-model II.

SUB-MODEL III

Sub-model III incorporates another type of SDP, the phantom midge, *Chaoborus*, which is a very important component of the plankton in many lakes. The model makes two assumptions: (1) planktivores will select *Chaoborus* over cladocerans and will take cladocerans over nearly equal or slightly larger-sized copepods; (2) *Chaoborus* will take copepods over nearly equal or slightly larger-sized cladocerans. Both of these assumptions have been discussed previously. The model considers cladocerans and copepods as well as *Chaoborus* and illustrates this application using a species pool from the western United States (table 6).

State 0

In the absence of GLPs, *Chaoborus* reaches its highest levels, preying primarily on copepods and constraining the development of their populations at State +1. The cladoceran prey will be the largest available, which, in this region of the world, is *D. middendorfiana*, maturing at sizes of 2.0–3.0 mm.

STATE + 1

Increasing planktivore predation will fall most heavily on *Chaoborus*, although the large daphnids will be affected and we anticipate vertical migration behavior at this time by both daphnids and *Chaoborus*. The *Chaoborus* population still feeds most heavily on young copepods, limiting these invertebrate predator populations. As GLP pressures increase, the abundances of *Chaoborus* decrease; as a result, the copepod populations should increase slightly. This state also corresponds to a change from the large *D. middendorffiana* to *D. rosea*, which matures at sizes less than 2.0 mm but is still able to produce adults of a size large enough to escape predation from the SDPs.

STATE + 2

As fish predation increases, the *Chaoborus* populations decrease further. There is a concommitant increase in pressures on the large daphnids, leading to a replacement by the smaller *D. schodleri*, which matures at sizes of 1.2 to 2.0 mm. Because the copepod populations are still controlled by *Chaoborus*, their effects on the daphnids are minimal; helmeted daphnids are not expected, although *D. schodleri* does have this ability. In those lakes where copepod pressures are significant, *D. schodleri* should be helmeted.

STATE + 3

Because of fish predation, few *Chaoborus* occur, allowing a full development of the predatory copepods. Given the increasing pressures on the daphnids from both fishes and copepods, *D. schodleri* should be replaced by the smaller *D. galeata mendotae* or other strongly helmeted species with a greatly reduced overall visibility to cope with these two pressures.

STATE + 4

With fish pressures increasing, a decrease should occur in the levels of predatory copepods, and daphnids should be replaced by the smaller-sized *Ceriodaphnia*. The predatory copepods present should be the smaller *Diaptomus* or *Cyclops*, rather than the larger *Epischura*.

State + 5 to + 6

As fish predation moves to its most intense level there is a complete absence of all SDPs with the exception of the rotifer *Asplanchna*, which exhibits increased transparency and is small enought to avoid heavy fish predation. This opens a niche for the smallest cladocerans, such Bosminidae as *Bosmina longirostris*.

As a result of the inclusion of *Chaoborus*, the main difference between Sub-model III and Sub-models I and II is the reduction of copepod predation in the first two states because of the *Chaoborus* preference for copepods over cladocerans. This allows the presence of three niches for daphnids without helmet development. The copepod populations never attain densities such as those in lakes without *Chaoborus* populations.

7. Testing the Models

To test these theoretical models of freshwater community structure, one needs a situation with different levels of GLP intensity in order to determine empirically whether the resulting zooplankton composition corresponds to that predicted. There are three possible cases:

1. A single lake where predation intensity differs over time, for example, where fishes have been either added or removed.
2. A single lake where predation intensity differs spatially, for example, where fishes are restricted to distinct littoral or pelagic regions.
3. A series of lakes differing in predation intensity, for example, having different densities or species of fishes.

The following section presents in detail three of the most thoroughly studied examples from the scientific literature that lend themselves to a rigorous test of the models.

MICHIGAN LAKES

During the years 1955 through 1965, Merle G. Galbraith, Jr., of the Institute for Fisheries Research, Michigan Department of Conservation, in Ann Arbor, Michigan, studied the survival of rainbow trout introduced to several Michigan localities and reported his results for Stager Lake and Sporley Lake (1967). The several fish species present in these two lakes since at least 1955 included pumpkinseed, bluegill, smallmouth bass, largemouth bass, and yellow perch. In 1958 rainbow trout were introduced to Stager

Lake and were restocked each subsequent fall. Fish stomach-content analysis indicated that the trout's only zooplankton food consisted of the genus *Daphnia*, and Galbraith decided to evaluate the effects on the daphnid prey population by sampling the plankton during the months of July, August, and September, when fish predation was most intense (i.e., the number of fish was at a seasonal maximum).

Zooplankton samples from Stager Lake showed a relatively stable situation during the initial period of sampling, 1958–60, with *D. galeata mendotae* and *D. retrocurva* being the predominant cladocerans encountered (some *D. pulex* appeared in 1959). *D. galeata mendotae* and *D. retrocurva* remained the dominant species through the summer of 1960, which was the final year of Galbraith's sampling for Stager Lake.

Sporley Lake presented a very different situation. In 1955 this lake received a treatment of Toxaphene in order to kill all fish prior to restocking. This chemical remained toxic to fish until 1959, when rainbow trout were successfully introduced, and this stocking was continued each year thereafter. Other introduced fish included fathead minnows in 1959 and smelt (which apparently entered illegally) in 1962. At first sampling, in 1956, Galbraith found that *D. pulex* was the only daphnid in Sporley Lake. No other species was found until 1960 when a few *D. retrocurva* appeared in the lake populations.

In the next four years Sporley Lake witnessed the stocking of rainbow trout but also, according to Galbraith, the buildup of large populations of minnows and smelt. The final years of study, 1964 and 1965, saw dramatic changes in the structure of the zooplankton community. The following conclusions from this study agree well with the model. With increasing gape-limited predation

1. The mean mature instar size of *Daphnia* decreases.
2. Smaller daphnids, which can mature at smaller sizes and are thus less vulnerable to fish predation, replace larger species.
3. The small daphnid species are replaced by still smaller cladocerans of other genera, such as *Bosmina*.
4. The predation intensity levels of SDPs increase, at least initially.
5. The appearance of helmeted daphnids coincides with a buildup of both GLP and SDP intensity.

In terms of the model, Stager Lake had remained basically in State +2, whereas Sporley Lake moved from State zero to State +2 and finally toward State +5 or +6.

CONNECTICUT LAKES

Galbraith's study compared changes in fish predation intensity and the resulting zooplankton community within one lake. The following example compares several lakes having different levels of predation intensity. In the study by John L. Brooks and Stanley I. Dodson of Yale University, the authors found that lakes in the Connecticut region appeared to fall into one of two categories with regard to open-water zooplankton composition. One lake type was characterized by the dominance of a single large *Daphnia* species and the presence of such large invertebrate predators as the cladoceran *Leptodora*, the copepod *Epischura nordenskioldi*, or both. This plankton community was associated with lakes in which there was an absence of the planktivore alewife, *Alosa*. Although other planktivores were present in these lakes, the alewife is an extremely efficient plankton-feeding species and was recognized as the most important predator in the region. The other type of lake had naturally landlocked populations of *Alosa pseudocharengus*. The plankton community in these lakes was characterized by small zooplankton, including the cladoceran *Ceriodaphnia*, but was dominated principally by *Bosmina longirostris*; relatively lower levels of the smaller herbivorous copepods; and small invertebrate predators absent from non-alewife lakes, including cyclopoids and also the predatory rotifer *Asplanchna*. In these alewife lakes there was a complete absence of *Daphnia*, *Leptodora*, and *Epischura*.

The hypothesis proposed by Brooks and Dodson (1965) was that the alewife produced a dramatic shift of the zooplankton populations, from one characterized by large-sized species in the absence of *Alosa* to one dominated by small-sized species in the presence of *Alosa*. To test this hypothesis they examined Crystal Lake, where a population of the closely related *Alosa aestivalis* had been introduced in the early 1950s and had since become abundant. Because Brooks had collected plankton from Crystal Lake in 1942, previous to the introduction, they were able to compare those samples with samples from 1964, ten years after the fish had become numerous.

The results supported their hypothesis. The 1942 Crystal Lake zooplankton populations, with *D. catawba, Leptodora*, and *Epischura*, resembled the modern zooplankton populations in non-alewife lakes. The 1964 populations, dominated by small forms such as *Bosmina* and *Ceriodaphnia* as well as small predatory copepods and the rotifer *Asplanchna priodonta*, resembled the natural Connecticut alewife lakes. The modal body-sizes of the Crystal Lake populations decreased from 0.785 mm without alewives to 0.285 mm with alewives present (see fig. 8). According to the authors, animals with a large mean size of first mature instar (including *Epischura*, about 1.6 mm; *Mesocyclops*, about 1.4 mm; *Daphnia*, 1.3 mm; and *Diaptomus minutus*, about 0.8 mm) were cropped by the fish to sizes below these minimum levels, resulting in local extermination. These species were replaced by zooplankton whose mean size of smallest mature instar was below 0.6 mm—*Ceriodaphnia*, about 0.5 mm; *Tropocyclops*, just below 0.5 mm; *Asplanchna*, 0.4 mm; and *Bosmina*, 0.3 mm. The one exception was *Cyclops bicuspidatus thomasi*, whose adult value of about 0.9 mm should have excluded it. The authors comment that adults of this species in general are restricted to the littoral areas of the lake where *Alosa* does not feed, whereas the smaller cyclopoid immatures are planktonic. Thus *Alosa* was probably encountering only the smaller-sized individuals and for this reason did not significantly affect the population.

The following conclusions from this study relate well to the model. With increased predation by GLP:

1. The mean mature instar size of large zooplankton prey decreases until no mature individuals are encountered.
2. Small cladocerans of other genera such as *Ceriodaphnia* or *Bosmina* replace the larger daphnid species.
3. The large SDPs are replaced by smaller predatory species, resulting in a reduction of SDP innensity.
4. The larger herbivorous copepods are replaced by smaller ones that are less heavily preyed upon by fishes and occur when existing SDP intensity is reduced.
5. The abundance of the smallest SDPs, such as the rotifer *Asplanchna*, coincides with an increase of small-sized cladoceran prey

The formerly dominant *D. pulex* was completely replaced by *D. retrocurva* and *D. galeata mendotae*. In addition, individuals of the smallest limnetic cladoceran, *Bosmina longirostris*, began appearing in measurable numbers, maturing at sizes well below the 0.8 mm minimum for daphnids.

An indication of the increase in predation intensity from 1958 to 1965 can be seen from the changes in average body-size of Sporley Lake daphnids. The mean value dropped from the 1.4 mm in the initial years of 1958–60 to a low of 0.8 mm by 1964–65. In addition, whereas the smallest mature instar found in Sporley Lake in 1960 was 1.2 mm, in 1964 this

TABLE 7. Chronological Changes for Fish and Zooplankton Species Composition in Two Michigan Lakes

	Fish	Cladoceran
Stager Lake		
1955	*Lepomis gibbosus*	*Daphnia retrocurva*
	L. machrochirus	*D. galeata mendotae*
	Micropterus dolomieu	
	M. salmoides	
	Perca flavescens	
1958–60	*Salmo gairdneri*	
Sporley Lake		
1955	Toxophene treatment	
1956	No fish	*D. pulex*
1959	*Salmo gairdneri*	*D. pulex*
1960	*Pimephales promelas*	*D. pulex*
1962	*Osmerus mordax*	
1964–65*		*D. retrocurva*
		D. galeata mendotae
		Bosmina longirostris

SOURCE: Galbraith 1967.
*Years of maximum *Epischura* abundance and the first appearance of *Leptodora*.

had decreased to 0.8 mm. This measurement was in striking contrast to values for Stager Lake, the control for the study, where no planktivore additions occurred after 1955. In Stager Lake the average size of *D. pulex* (briefly present) at maturity was 1.8 mm, with some *D. galeata mendotae* and *D. retrocurva* reproducing at sizes of up to 1.4 mm and 1.2 mm respectively. Unfortunately, Galbraith did not record numbers of SDP predators in Sporley Lake but noted that *Leptodora* first appeared during the last years of his sampling (pers. comm., cited in Dodson 1970). In addition, subsequent examination of his lake samples indicated that the predatory copepod *Epischura* was rarely present or entirely absent from plankton samples in Sporley Lake during the years through 1960. After this year, coinciding with the increase in planktivore numbers, *Epischura* first appeared, and the population reached its maximum abundance in 1964–65 (Kerfoot, pers. comm.). The complete changes in the Michigan lakes are presented in table 7.

Relating this to Sub-model I (Table 8), in the absence of alewives (but with some fish species present), the lake is in State +1, with *D. catawba* dominant. With the introduction of alewives there is a transition to State +4, with *Ceriodaphnia* present, and eventually to state +6, char-

TABLE 8. Percent Composition of Zooplankton Species in Connecticut Lakes with and without Alewives

	Lakes Without Alewives (State +1 +2)	*Lakes With Alewives (State +4 +5)*
Cladoceran prey		
Daphnia catawba	5.5	0
Daphnia galeata	15	0
Ceriodaphnia lacustris	0	14
Eubosmina coregoni	4	0
Eubosmina tubicens	1	1
Bosmina longirostris	1	27
Holopedium sp.	2.5	1
Diaphanosoma sp.	2	2
Size-dependent predators		
Leptodora kindti	1	0
Epischura nordenskioldi	4	0
Cyclops bicuspidatus	2	17
Mesocyclops edax	6.5	0
Orthocyclops modestus	0	4
Tropocyclops prasinus	0	11
Asplanchna priodonta	0	7
Copepod prey		
Diaptomus minutus	40	0
Diaptomus pygmaeus	13.5	5

SOURCE: Brooks and Dodson 1965.

acterized by *Bosmina* dominance. Considering also Sub-model II, this transition is accompanied by the disappearance of such large invertebrate predators as *Epischura* and *Leptodora* and the reduction of such large herbivorous copepods as *Diaptomus* and their replacement by smaller predatory copepods. *Asplanchna* becomes an important component of the plankton.

NORTHERN SWEDISH LAKES

Another test of these models is available from studies at the Institute of Freshwater Research in Drottningholm, Sweden, by N. A. Nilsson and B. Pejler, whose examination of northern Swedish lakes asked, "Is there a correlation between fish fauna and zooplankton composition?" Their recent work (Nilsson and Pejler 1973) considered twenty-eight lakes and also included data from several other workers, comparing information on the fish fauna with data from zooplankton collections. The lakes range from those having no fish up to those having three plankton-feeding species. The fish include the brown trout, *Salmo trutta*, which was commonly introduced into barren Swedish lakes and shows some planktivory, especially in its larvae or juvenile stages. When present with other fish species, *S. trutta* relies more heavily on benthic prey. The effect of brown trout on limnetic plankton populations, therefore, is minimal. The arctic char, *Salvelinus alpinus*, had also been introduced into otherwise barren lakes, and this species is planktivorous throughout its life, attaining sizes of 40 cm or more. Char and brown trout are known to compete for benthic animals as food when introduced together in a lake (Nilsson 1960, 1963). Char can consume smaller items than can brown trout, and thus we would expect this fish to be a more efficient planktivore than *S. trutta*. A third group of fish in many of these lakes includes the genus *Coregonus*, the whitefish, individuals of which are able to consume plankton particles far smaller than those eaten by either *S. trutta* or *S. alpinus*. This difference in feeding efficiency is used by the authors to explain why char often disappear from lakes following the introduction of whitefish. Most lakes with whitefish may have two or more *Coregonus* species, but because the taxonomy is still uncertain (at times they may be hybrids rather than distinct species), the designation *Coregonus* is used without specific names.

These three planktivores present an opportunity to compare different lakes under different levels of predation intensity. The lakes can be classified as five types. Type zero lakes include "fishless" lakes (although the lake may contain fish, it contains no planktivores). Type I lakes are those in which only trout have been introduced. Type II lakes have only the arctic char or arctic char and trout.* Type III lakes have char, trout, and whitefish. Type IV lakes contain whitefish or whitefish and trout.

The authors assume that predation intensity increases in these lakes from a minimal level to the most intense level of planktivory. As an independent check of this assumption, one can examine the mean body-size of the genus *Daphnia* among these lake types. The actual values are: greater than 2.5 mm (type zero); 2.34 mm (Type I); 1.68 (Type II); and 1.08 (Type III). There are insufficient daphnids to evaluate Type IV. This decrease in *Daphnia* population mean body-size provides strong support for the as-

* In the original publication the arctic char and trout combination is considered a separate type; but according to Nilsson and Pejler, the trout may have no effect on the plankton in this combination because of the competition from char, and therefore I have categorized these lakes as Type II.

sumption that Types zero to IV represent a pattern of increasing predation intensity by planktivorous fishes. The zooplankton associations are consistent within lakes of each type, indicating a correlation with levels of SDP intensity. The authors demonstrate that these prey assemblages do not correlate with any measured physical parameters (other than location in the north Swedish forest or mountain region), so this can be eliminated as a possibility.

Quantitative data are lacking because the plankton hauls were vertical tows only. Instead, the authors present qualitative evaluations for each species: (1) "absent" (0), (2) present (+), (3) common (+ +), and (4) abundant (+ + +). Samples were taken from 1960 to 1971, but approximately 75 percent of the samples of the twenty-eight lakes are from 1968–71.

All zooplankton data given by the authors are used here (table 9), with the exception of the copepod *Mixodiaptomus laciniatus*, which was rare; the copepod *Heterocope borealis*, which appeared only once; and the *Diaptomus* and *Cyclops* categories, which were composed mainly of immatures.

The following generalizations are given for the "characteristic species," which are those occurring in at least 50 percent of the lakes sampled.

Polyartemia forcipata is the largest crustacean, and although it was found only once in the twenty-eight lakes studied, the authors tell us that this species was formerly characteristic in fishless lakes, such as Lake Piekejaure. After the introduction of arctic char, however, the *Polyartemia* population, found to be the main food of this fish, was eliminated. This and other evidence leads the authors to conclude that *Polyartemia* is a main prey for planktivorous fish and exists only in fishless lakes (Type 0).

Of the four daphnid species, *D. longispina* is by far the largest, attaining lengths of almost 3.0 mm. Next, in decreasing order, is *D. longiremis*, *D. galeata*, and *D. cristata*, the last maturing at sizes below 1.0 mm. Although there is at least one daphnid species in all lake types, the particular species and its abundance vary according to the level of fish predation intensity. *Ceriodaphnia quadrangula* and *Bosmina longirostris*, the smallest cladocerans, appear commonly only in Types III and IV.

The dominant SDPs are *Bythotrephes longimanus* and the large calanoid copepod genus *Heterocope* (two species), with *Leptodora kindti* being less common. These species are most abundant in lakes of intermediate fish predation. There is no mention of *Chaoborus*, although it may have been present in the fishless lakes.

Among the remaining herbivorous species, *Holopedium* is abundant

TABLE 9. Abundance and Likelihood of Occurrence of Zooplankton Species in Northern Scandinavian Lakes Differing in Fish Species Composition

	Type 0 Fishless		Type I Trout		Type II Char/Trout		Type III Char/Trout Whitefish		Type IV Whitefish Trout	
Number of Lakes N =	5	11	8	10	9	29	2	5	4	10
Polyartemia forcipata	present		0	0	0	0	0	0	0	0
Daphnia longispina	63	45	37	70	0	7	0		0	
Daphnia galeata	0	0	13	20	29	90	36	80	22	50
Daphnia longiremis	0	0	0	0	0	14	0	0	0	0
Daphnia cristata	0	0	0	0	0	3	0	40	100	70
Ceriodaphnia quadrangula	0	0	0	0	0	0	40	40	60	50
Eubosmina coregoni	7	55	31	100	32	100	18	100	12	100
Bosmina longirotris	0	0	0	0	0	0	0	0	100	30
Bythotrephes longimanus	0	0	86	50	0	48	14	40	0	30
Leptodora kindti	0	0	0	0	0	10	0	60	0	40
Heterocope saliens	0	0	79	60	21	59	0	20	0	0
Heterocope appendiculata	0	0	0	0	0	0	100	80	0	40
Holopedium gibberum	4	18	29	80	28	86	19	100	20	70
Eudiaptomus graciloides	0	9	0	0	17	24	59	100	24	70
Acanthodiaptomus denticornis	0	0	0	0	0	17	0	0	100	20
Arctodiaptomus laticeps	23	27	31	50	12	69	28	80	6	30

SOURCE: Nilsson and Pejler 1973.

NOTES: *The average abundance per lake*. This value is a computed average for all lakes in this category, using Nilsson and Pejler's qualitative assessments of species' abundance. If, for example, a species were in only two of the five fishless lakes, common in one (+ +) and abundant in the other (+ + +), it would be assigned a final value by taking the sum of these pluses and dividing by five. This value expresses the average abundance of species in this lake type. Values are taken from the authors' Table 2.

The percentage of lake types including a given species. This second value expresses the probability of encountering a species in the particular lake type. Values are taken from the authors' Table 3, which includes data from previous workers as well from their own study.

in all but the fishless lakes. The largest herbivorous copepod, *Arctodiaptomus laticeps*, which attains lengths of almost 2.0 mm, is most abundant in the intermediate type lakes and appears to be replaced in Type III lakes by *Acantodiaptomus denticornis*, with which it is never sympatric (Nilsson and Pejler 1973). From the information presented in table 9, one can assemble a model for the northern Swedish lakes (table 10) which summarizes the trends for each species over the five lake types.

STATE 0

The fishless lakes (Type Zero) are dominated by *P. forcipata* and *D. longispina*. The large herbivorous copepod *A. laticeps* is present, but there are no SDPs, thus classifying these lakes as State 0 for them also.

STATE + 1

When trout are present (Type I), the largest zooplankters, *P. forcipata*, are absent because of predation, and the largest daphnid, *D. longispina* is dominant. The large herbivore *A. laticeps* is abundant, as is the small cladoceran *E. coregoni*. The daphnid *D. galeata*, whose normal mature size range is 0.8–1.8 mm, is common also, as is *H. gibberum*. The increase in abundance of smaller-sized prey coincides with the first establishment of an SDP population in the form of *Bythotrephes*. The large size of this animal, up to 5.0 mm, is tempered by its relative transparency, which confers some degree of protection from fish predation. Also present is the large (up to 3.0 mm) predatory calanoid copepod *Heterocope saliens*, which attains its maximum abundance in these lakes. Given the presence of both fish and invertebrate predators, we anticipate the presence of *D. galeata* and *Holopedium*, both of which have reduced visibility and also morphological adaptations to protect them from invertebrate predators. The former has a helmet development; the latter has a gelatinous matrix, which may make it less desirable to fish. With its relatively large size (1.5–2.5 mm) and matrix, *Holopedium* is protected from invertebrate predation (see chapter 4).

STATE + 2

As predation intensity is increased by the dominance of char (Type II), *D. longispina* is effectively replaced by the smaller *D. galeata* and the helmeted *D. longiremis*, which occurs only in these lakes. These two helmeted cladocerans are the dominant prey types. The smaller *D. cristata* occurs, although this species remains rare. The same species of large in-

TABLE 10. Zooplankton Community Model for Northern Swedish Lakes

State	*Gape-Limited Predator*	*Cladoceran Prey*	*Other Prey*	*Size-dependent Predator*	*State*
0	Fishless (0)	*P. forcipata* *D. longispina* *E. coregoni*	(*A. laticeps*)	None	0
+1	Trout (I)	*D. longispina* *E. coregoni* (*D. galeata*)	*A. laticeps* *H. gibberum*	*H. saliens* *B. longimanus*	+2
+2	Char/Trout (II)	*D. galeata* *E. coregoni*	*H. gibberum* (*A. laticeps*) (*E. graciloides*)	(*H. saliens*) (*B. longimanus*)	+2
+3	Char/Trout/ Whitefish (III)	*D. galeata* *D. cristata* *C. quadrangula* (*E. coregoni*)	*A. laticeps* *E. graciloides* (*H. gibberum*)	*L. kindti* *H. appendiculata* (*H. saliens*) (*B. longimanus*)	+4
+4	Whitefish/Trout (IV)	*D. cristata* *C. quadrangula* (*D. galeata*) *B. longirostris* (*E. coregoni*)	*A. denticornis* (*A. laticeps*) (*H. gibberum*) (*E. graciloides*)	*L. kindti* (*B. longimanus*) (*H. appendicu- lata*)	+2

SOURCE: Nilsson and Pejler 1973.

NOTE: The vertical columns from left to right give (1) the state of the GLP level; (2) the lake type; (3) the cladoceran prey associated with that state; (4) other herbivorous crustaceans, mainly copepods; (5) the SDPs; and (6) the state of predation intensity for the SDPs. Species listed in parentheses are those whose abundance in that particular category is less than 25 percent of their total population distribution in each of the five lake categories.

vertebrate predators still are not affected appreciably by fish predation and their numbers remain common (State +2). The herbivorous copepod *A. laticeps* is common, and in addition the smaller herbivorous copepod *Eudiaptomus graciloides* is present.

State + 3

With the addition of whitefish (Type III), *D. galeata* is still the dominant daphnid; also present are *D. cristata* and the even smaller *Ceriodaphnia*. The smaller cladocerans can persist because both of the large SDPs, *Heterocope* and *Bythotrephes*, are being reduced by increased fish predation. The transparent *Leptodora* dominates, and the predatory *H. saliens* is replaced by the smaller congeneric *H. appendiculata*. As a result, both herbivorous dominants, *Holopedium* and *Eudiaptomus*, are found in 100 percent of Type III lakes; and *A. laticeps* also increases slightly in abundance. This is the level of the most intense size-dependent predation because there are four different invertebrate species occurring simultaneously.

State + 4 to 5

In lakes dominated by whitefish (Type IV), the intensity of GLP reaches its maximum, which in this system corresponds to State +4. There has been a change in the daphnids, with dominance by *D. cristata* (present in seven of the ten Type IV lakes). *Ceriodaphnia* is also common (present in five). *Bosmina* appears for the first time, occurring in three of these lakes (perhaps State +5). Among the invertebrate predators, the levels of both *Leptodora* and *Heterocope appendiculata* start to decrease, both occurring in only four of these lakes. The total SDP level is decreased, to State +2, which helps to explain the dominance of the smaller cladocerans. Regarding the herbivorous crustaceans, the large *A. laticeps* is in decline, being replaced by the smaller herbivore *Acanthodiaptomis denticornis* for the first time. E. *graciloides* is now present in seven of the ten lakes, as is *Holopedium*.

In this study certain predicted patterns occur.

1. With very low GLP intensity, the zooplankton community is dominated by large herbivores.
2. With some GLP pressures there is a replacement of large daphnid species by smaller forms until the smaller herbivore genera become the dominants.
3. With further increases in GLP intensity, an initial increase in the

dominance of large SDPs occurs as their prey density increases, which shifts to a dominance by smaller or less conspicuous invertebrate predators and finally to a decrease in the large SDPs as their numbers are reduced by GLP.

4. When SDP pressures are greatest, cladoceran herbivores dominate. These cladocerans—for example, the helmeted daphnids and *Holopedium*—are able to cope with both types of predators.
5. With increasing GLP intensity and the reduction in numbers of large-sized SDPs, large-sized herbivorous copepods are replaced by smaller species, which can persist once the largest invertebrate predators are in decline or absent.

Given our knowledge of the local Swedish fauna, the community models allow us to predict the species composition of any lake in this region when the fishes present are known. The single exception is the species *E. coregoni*, which occurs, inexplicably, in all lake types.

To test the explanatory power of these models, I selected three comprehensive studies that included predator and prey data. There are, however, innumerable examples in the literature that could be presented in an equally convincing analysis, which the reader might use for his or her own test. The main conclusion from these tests is that the models, based on interactive predation, give reasonably accurate predictions of lake species composition of zooplankton communities. This shows clearly that in freshwater communities the effect of SDP predation is a decisive, often overwhelming, selective force. By understanding predation, one can make very broad predictions about the entire lake community. More specifically, the gape-limited predator is the most important controlling component (the best predictor) of the zooplankton community structure.

The model has three basic states:

State 0 to + 1 is characterized by:

1. A complete absence or low levels of GLPs.
2. Low levels of SDPs.
3. Prey species of large body-size.

State + 2 to + 3 is characterized by:

1. Moderate levels of GLPs.
2. High levels of SDPs.

3. Prey species of medium size with such morphological adaptations as helmets, tailspines, or other means of physical protection.

State + 4 to + 6 is characterized by:

1. High levels of GLPs.
2. Low levels of SDPs.
3. Prey species of small size.

Given these three general states for a particular lake, the specific GLP and SDP will determine the number of intermediate states.

8. The Role of Competition

The models developed in the preceding chapters allow us to predict prey types by knowing only which predators are present in the system. Although in some cases it may be possible to identify prey at the species level, more usually it is possible only to specify general prey types. For instance, in the absence of GLPs we predict a large herbivore. Whether this will be a cladoceran or a copepod and which genera will prevail remain to be seen by investigation. The same kind of question can be asked when both GLPs and SDPs are present because we predict a prey that can persist under the constraints of both predator pressures. But we would not know whether to expect a helmeted daphnid, two species, or a guild of copepods. Predictions can be improved at the species level by looking at competition, and this chapter will examine the role of competitive interactions in determining zooplankton community composition.

Competition among species occurs in every community where required resources are in limited supply. Although a great deal of ecological theory concerns competition (see MacArthur 1972, Cody 1974, for reviews, discussion, and literature references), as Hutchinson noted, one is faced with "the extreme difficulty of identifying competition as a process actually occurring in nature" (1957, p. 418). Competitive interactions between two individuals may be obvious and easily witnessed, but usually the effects of competition can be seen only over an extended period of time, often longer than the observation period of the investigator. This characteristic of competition leads to certain intractable problems when attempting field measurements of competition (see Orians and Horn 1969, Vandermeer 1970, or Zaret and Rand 1971 for discussions). As a consequence, nearly all field studies of competition have relied heavily on correlative or other indirect evidence of competition. Few field studies of competition have been able to adopt experimental methodology, including controls, which critics such as Connell (1975) and others feel is necessary. It is probably because of this overall lack of well-supported, direct evidence that competition is not emphasized as an important determinant of freshwater community structure, in spite of its being proposed as a critical factor in one of the first papers to consider determinants of zooplankton community composition (Brooks and Dodson 1965). Although some field experiments have suggested that competition is, at best, a weak force in freshwater communities (Dodson et al. 1976), a better conclusion is that its role has not been investigated with sufficient scientific rigor. This chapter will explore ways to elucidate the role of competition in freshwater communities.

When we use predation models for predicting zooplankton species composition what we are really predicting is gross morphology, including prey body-size, body-shape, pigmentation, and to some extent behavior (e.g., vertical migration). Since taxonomists generally use gross morphology for species identifications, this results in surprisingly accurate species or generic predictions. Competitive relationships determine which of the several prey species that can potentially fill this role will be the dominant prey organisms. In evolutionary terms we would say that if predation determines that aspect of the prey-realized niche (chapter 9) called gross morphology, competition should determine those aspects called fine morphology, especially feeding morphology. The following sections illustrate how competition and predation interact in resolving zooplankton species composition at a finer level.

HERBIVOROUS PREY

Given known predation levels for both GLPs and SDPs, one can predict prey types. Predicting the dominant species, however, is another matter. For instance, if one predicts a helmeted daphnid, which of the several available ones will appear in the lake? Or if one observes a niche for a small herbivorous copepod, will it be *Diaptomus* or *Eucyclops*, and which species in the genus? The answers depend on which animal is the best competitor in the particular situation.

Of the two morphs of *Ceriodaphnia cornuta*, one has an advantage in the presence of GLPs. The other morph is the competitive superior and is dominant in the absence of planktivorous fish (Zaret 1972*a*). Many of the polymorphisms among limnetic animal populations (see review in Hutchinson 1967) are a result of predation pressures, and it has been suggested that the morph with the lower predation rate will always suffer in reproductive ability (Zaret 1972*b*). The very adaptation that lowers its susceptibility to predation may also reduce its food-gathering abilities.

A recent study investigated the two morphs of *Bosmina longirostris* that are found in Lake Washington (Kerfoot 1977*b*). Using *in situ* enclosures with the two *Bosmina* morphs, Kerfoot showed that the short-featured morph is the superior competitor, a result mainly of its production of more eggs per clutch, a greater total yolk content per brood, and its general ability to mobilize food resources rapidly into reproduction during times of abundant food. Although under these conditions the long-featured morph is the inferior competitor, its reduced rate of predation from the copepod *Epischura*, allows it to be the dominant morph when copepod predation is high (Kerfoot 1977*a*).

The smaller visible carapace and reduced eye of helmeted daphnids allow them a certain inconspicuousness and consequently they are less likely to be seen by planktivorous fish than the nonhelmeted morphs. These same helmeted animals have a reduced motion component (Jacobs 1967), which also makes them less "visible." Reduced motion may mean that their filtering rate is reduced, making them inferior competitors with nonhelmeted morphs.

The first serious consideration of competition among freshwater zooplankton concerned the relative food-gathering abilities of different-sized crustaceans. The "size-efficiency hypothesis," proposed by Brooks and Dodson (1965), stated that whereas small filter-feeding herbivores were limited to a fairly narrow food-particle range, larger species had an exploitable particle-size range that overlapped that of the smaller species. This implied that larger species would be better competitors.

This hypothesis was supported by several laboratory studies of filtering by herbivorous zooplankton. Burns (1968*a*, 1969*a*) fed different cladoceran species micronic beads (minute plastic spheres) and bacteria ranging in size from less than 1 up to 80 μ. She found a strong positive correlation between carapace size and the largest particle that could be ingested. Gliwicz (1969) used sand grains as filterable particles for his zooplankton in the laboratory. He also demonstrated that the available size range of particles increased with herbivore size (fig. 31). Work on the feeding mechanism of the planktonic marine copepod *Calanus pacificus* by Frost (1972,

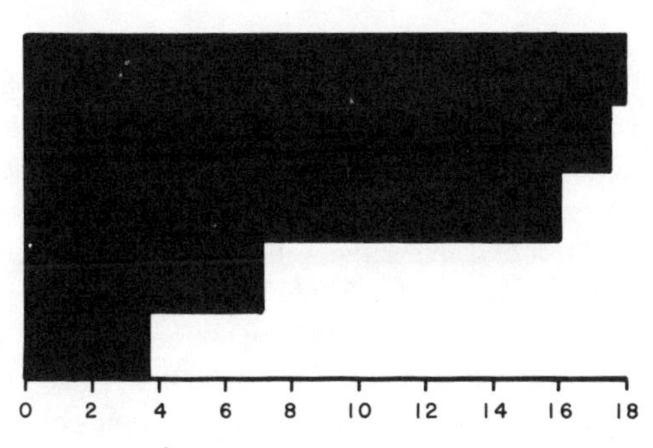

Fig. 31. Size ranges (in microns) of ingested particles for different zooplankton herbivores. (From Gliwicz 1969)

1974, 1977) supported the view that copepods also are indiscriminate filter feeders, although they are more efficient at capturing, handling, and ingesting the large particles. These studies suggested that zooplankton were passively selecting organisms, ingesting particles simply as a function of available algae and the distance between the filtering setae of the animal. The concept of zooplankton herbivores as exhibiting "passive selectivity" appeared to support the size-efficiency hypothesis, as shown *in situ* grazing experiments by Haney (1973) in Heart Lake, Ontario. Haney showed that the average filtering rate correlated well with carapace length, the two largest daphnids accounting for almost 80 percent of the total annual lake value. This suggested that in the world of passively selective zooplankton herbivores, the largest sizes would be the best competitors. The size-efficiency hypothesis had a great deal of empirical support.

Other work, however, has disputed the validity of these conclusions from feeding experiments. Although the area of filtering setae increases in daphnids approximately as the square of the body-length, Egloff and Palmer (1971) found that the filtering rates of *D. rosea* became greater as body-length increased compared with the larger *D. magna*. Perhaps the most startling results were those of Wilson (1973). He interpreted his data for the marine copepod *Acartia tonsa* to mean that *A. tonsa* was able to shift to algae particle sizes as they became most abundant in its environment, indicating that it was a rather selective feeder. (See other viewpoints and critiques of Wilson in Boyd 1976 and Frost 1977. Recent work has considerably strengthened this point of view, however.)

Other laboratory and field data seemed to contradict the size-efficiency hypothesis. Earlier, Frank (1957) had demonstrated that the large

D. magna was often outcompeted by the smaller *D. pulex* in laboratory studies. Of the two *Ceriodaphnia cornuta* morphs from Panama, one form was definitely superior competitively even though both morphs were the same in body-size (Zaret 1972*b*). It is the larger morph of the cladoceran *Bosmina longirostris* that is the inferior competitor (Kerfoot 1977*b*). Neill (1975) found that large animals were less efficient at filtering small food particles and that their juvenile stages might actually be outcompeted by adults of other species feeding on food particles of the same size ranges. A review of the size-efficiency hypothesis (Hall et al. 1976) concluded that although it was possible there was insufficient evidence to strongly support its implications.

As information about relative competitive ability was considered, the importance of size-dependent predation in structuring aquatic communities continued to accumulate. As evidence mounted for the selective removal of smaller-sized prey by size-dependent predation, another explanation for the dominance of large-sized zooplankton in the absence of fish was proposed by several workers (Dodson 1974*b*; Zaret 1975; Kerfoot 1977*a*). Namely, in the absence of fish, SDPs would selectively remove any small prey species and leave only the larger ones. This provided an alternative hypothesis, based on predation factors, to account for community composition.

It is likely that the size-efficiency hypothesis is not accurate for several reasons. First, it assumes that competition among zooplankton occurs throughout the year (otherwise the small forms would return), especially in the spring and early summer when the larger zooplankton forms come to dominate. In fact, this is the time of year when competition is probably least important, because food is often very abundant and unlikely to be limiting (Porter 1976). Second, even if the particle range of large filtering species overlaps that of smaller forms, it does not necessarily mean that larger species are more efficient than small herbivores in the small-sized particle ranges (and see evidence in Neill 1975 that they are not). In addition, even if larger species have higher filtering rates, filtering efficiencies, or other advantages, smaller species may be more efficient in such aspects as incipient limiting food concentrations, ability to resist toxins, and others (pointed out by Hall et al. 1976).

Competition occurs between two species when a resource is limiting or below levels necessary for reproduction and growth. In none of the studies of food competition among zooplankton has it been demonstrated that food is a limiting factor. To date, this is one of the major criticisms of studies examining the effects of competition.

It is likely that different zooplankton species are optimally adapted for different conditions (see Lynch 1977) and that groupings of species occur in response to specific selective forces. When fish predation is present, there are several possible life-history patterns for prey species. One type is characterized by constant activity, a high reproductive rate, but small body-size. It includes *Ceriodaphnia* and *Bosmina*, species dominant only when gape-limited predation is intense. Another type is composed of species with a larger body-size and a concomitantly reduced motion component that makes them less conspicuous to fish and less heavily eaten by them, but as a result, they have a greatly decreased reproductive rate. This type, which includes some of the calanoid copepods in the genus *Diaptomus* and also such cladocerans as *Diaphanosoma*, occurs under moderate predation pressures. A final group includes those species that respond to the predation pressure of invertebrates with morphological adaptations but may have diminished competitive ability. A good example is *Holopedium*, whose gelatinous matrix makes it less preferred by fish, although its numbers fall sharply when more preferred food items are depleted (Stenson 1974; O'Brien 1975). The selective advantage of this external protection is apparent in the presence of heavy invertebrate predation to which *Holopedium* is relatively immune, but it is apparently a very poor competitor with other large cladocerans, so that *Holopedium* will be outcompeted by daphnids (Allan 1973).

GAPE-LIMITED PREDATORS

Several studies have examined the competitive relationship of different planktivores. In Swedish lakes the arctic char (*Salvelinus alpinus*) can feed on plankton throughout its life cycle, whereas only the larvae and juveniles of brown trout (*Salmo trutta*) exhibit planktivory. When trout alone are present in lakes, the young feed on plankton; when char are sympatric with them, there is a distinct habitat separation, with the char feeding almost exclusively on zooplankton and the trout restricting their diet to benthic food items (Nilsson 1960, 1963). This suggests that char will competitively exclude the apparently less-efficient plankton-feeding members of the trout population from the open waters of lakes. If whitefish (*Coregonus* sp) are introduced to lakes containing arctic char, the char eventually disappear, which leads to the conclusion that whitefish will outcompete char and cause their eventual extinction (Nilsson and Pejler 1973). These studies suggest a hierarchy of zooplankton-feeding efficiency among the three salmonid species.

Among North Temperate equivalents, the salmonid Dolly Varden (*Salvelinus malma*) is a superior competitor to cutthroat trout (*Salmo clarki clarki*) when feeding on benthos, but the reverse holds true when they are feeding on surface-associated foods (Schutz and Northcote 1972).

An important study on competitive interactions among potential Class I predators in the same water body is the recent work by Werner, Hall, and their associates from Michigan State University. These workers used experimental manipulation to examine competitive interactions among several centrarchids, including the bluegill, *Lepomus macrochirus*, the pumpkinseed, *L. gibbosus*, and the green sunfish, *L. cyanellus* (Werner and Hall 1976). When stocked in artificial ponds as single species populations, all three species preferred vegetation-associated insects and crustaceans (41 to 61 percent of their diet) over open-water zooplankton (1 to 8 percent) or benthic-associated insects (10 to 23 percent). The three artificial ponds represented controls in that each species was allowed to feed on whatever food it chose. When all three species were stocked in the same pond, however, bluegills increased their intake of zooplankton to 33 percent, and pumpkinseeds increased their consumption of benthic prey to 34 percent. Each species showed a drastic drop in feeding on the vegetation-associated food items (a drop to 15 percent and 5 percent respectively). The shift is illustrated in figure 32. The diet of the green sunfish showed little change when living with the other species, leading the authors to conclude that it was the competitive dominant, and that when all three species were present, the most "flexible" species, the bluegill, would become the dominant planktivore in the lake. Although there are other ways of interpreting these data (see Maiorana 1977 for criticisms), this approach is probably the first to examine the competitive interactions of fish using controlled experimental methods. It provides strong inferences for the direct effect of exploitative competition by one species on another and suggests that competition explains the habitat separation that occurs among these fishes in natural ponds in Michigan where the three species (as well as others) occur together (Werner et al. 1977; Werner and Hall 1977; Werner 1977).

The conclusion from these studies of sympatric fishes feeding on similar foods is that some species are more likely than others to be the domi-

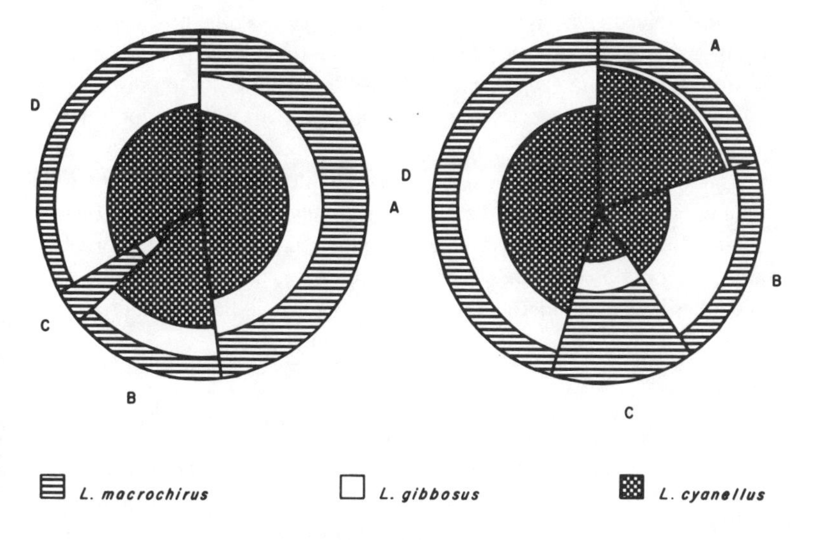

Fig. 32. Diet shifts in three species of sunfish for (*A*) vegetation-associated insects; (*B*) vegetation-associated crustaceans; (*C*) open-water zooplankton; (*D*) benthic-associated insects. (From Werner and Hall 1976)

nant planktivores. Studies of competitive interaction can help to predict which fish species will prevail and what the resultant level of GLP intensity will be in the final form of the zooplankton community.

SIZE-DEPENDENT PREDATORS

Information about competition between SDPs ideally would allow one to identify the species likely to be present given known GLP levels, but, unfortunately, little information is available on the competitive abilities of SDPs.

Intense food competition between predatory calanoid and cyclopoid copepods may result in temporal separations, that is, populations in the same lake with numerical dominance at different times of the year (Strickler and Twombly 1975). Predatory calanoid copepods may have evolved specifically for feeding on limnetic prey, which explains why they are usually dominant in open waters, whereas predatory cyclopoids are the important species in lake littoral zones.

The invertebrate predators with well-developed compound eyes probably rely on vision for prey capture (e.g., *Polyphemus, Leptodora, Mysis*), whereas others used tactile information exclusively (*Cyclops, Epischura, Chaoborus, Asplanchna*). Relating these attributes to prey capture ability presents some fascinating possibilities. If visual information is more efficient for capturing prey, predators would have to feed diurnally and their motion would make them more easily captured by fish during periods of light. Nonvisual invertebrate predators, on the other hand, might be less efficient but also less vulnerable to gape-limited predation. Field or laboratory competition experiments could develop this information, but to date no studies have examined the relative competitive abilities of SDPs.

COMMUNITY STUDIES

Several approaches have been used to examine the importance of competition in zooplankton communities. Dumont (1972) suggested that vertical migration patterns might allow the coexistence of competing filter feeders by spatial separation, and later this was examined in some detail by Lane (1975). Several other studies have suggested that food limitations in communities might result in change in species composition (Clark and Carter 1974; Kwik and Carter 1975) and have attempted to show that zooplankton species have sufficiently different niche parameters to avoid competing (Miracle 1974; Makarewicz and Likens 1975). Use of manipulative field techniques to examine food competition has been, unfortunately, rare, and the few existing studies have reached opposing conclusions (Sprules 1972; Allan 1974; Dodson 1974a; Dodson et al. 1976). Clearly, competition among zooplankton needs to be understood but because of technical and conceptual problems it is still in the early stages of investigation.

9. Interactive Predators

Freshwater communities are influenced not by one dominating species or class of predator but by several interacting predators, and the interaction can be either direct or indirect. Direct interaction occurs when one predator consumes another, thus affecting the abundance or distribution of the latter. For example, in freshwater communities planktivorous fish prey on the invertebrate predators *Chaoborus* or *Epischura*, controlling the level of Class II predation intensity. Indirect interaction occurs when the type of prey removed by one predator affects the abundance and size distribution of prey consumed by other predators. In alpine ponds, for example, the feeding by *Ambystoma* results in a skewed prey size-distribution that apparently allows the development of *Chaoborus* populations and thus results in Type A communities. When gape-limited predators are present in freshwater communities, one type of zooplankton community is favored; in their absence, the dominance of a size-dependent predator favors another species composition; when both are present a third type of community results. In all cases it is the interaction between dominant predators that determines the species composition of the communities.

Interactive predators appear to be a natural consequence of the predation process for several reasons. First, the most basic characteristic of predators is that they are highly selective. Numerous studies have indicated that predators preferentially remove the youngest and weakest members of prey populations (e.g., see review in Curio 1976), select individuals according to body-size (Schoener 1969), act in a "prudent" manner (Slobodkin 1961), or take the most conspicuous or isolated individuals when feeding on schools of prey (Edmunds 1974; Milinski and Curio 1975). Although scientists may argue about the feeding mechanisms of predators, they concur that predators are not taking random samples of the prey populations; rather, they are highly selective. Given that predators are discriminatory in their prey preferences, they are selective in ways that produce different consequences for the prey community. The electivity curve of fishes and salamanders increases monotomically with prey size; that of crustaceans and *Chaoborus* is bell-shaped. Each predator type has a distinctly different electivity curve, and each affects the prey community composition in different ways. Finally, there are a limited number of ways to detect and capture prey in freshwater limnetic zones (Zaret 1975). Limnetic habitats are unstructured, physically homogeneous areas with no hiding places for prey. The selective nature of predators, the different possible ways of being selective, and the limitations of the environment result in predators that interact with each other directly and indirectly through food resource removal (Armstrong and McGehee 1976; Lynch 1979).

Several examples of freshwater communities whose species composition is determined strongly by interactive predators have been discussed earlier in detail. To review, in New England, alewife lakes are characterized by distinct zooplankton assemblages, related to the type of prey selection by fishes. Large-sized, conspicuous prey are removed; small, less conspicuous forms survive. The skewing of the prey size distribution toward smaller body-sizes favors the presence of other predators such as copepods and rotifers, which are only found because of the interactive effects of fishes. In a second example, the alpine ponds of Colorado, the interactive predators are salamanders and copepods. When only predatory *Diaptomus* occur, the ponds are dominated by large-sized cladocerans and the fairy shrimp, *Branchinecta*. When larval *Ambystoma* appear, their interactive effects result in a new prey community.

The influence of interactive predators is not restricted only to limnetic regions. In Wisconsin lakes the littoral region is the scene for a different set of interactive predators. Smallmouth bass (*Micropterus dolomieui*), rock bass (*Ambloplites rupestris*), and yellow perch (*Perca flavescens*) feed heavily (usually more than 10 percent by dry weight) on the crayfish *Orconectes propinquus* (Stein 1977). Extensive field and laboratory experiments, the latter with *Micropterus* as the predator on *Orconectes*, have documented the relationship between the feeding preferences by the fish and the morphological and behavioral adaptations of the prey (Stein 1976; Stein and Magnuson 1976; Stein, Murphy, and Magnuson 1977). Crayfish abdomen widths (related to swimming speed), protective carapaces, and chelae size are three morphological traits that can reduce fish predation on individual crayfish. Differential susceptibility to predators—prey body-size, life stage (juveniles versus adults), state of maturity ("berried" females carry eggs), or increased vulnerability during intermoult periods—can be modified by the crayfish through such behavioral traits as substrate selection (microdistribution), a diurnal activity pattern, the use of shelters, and escape behavior (see Ware 1973). This evolutionary "accommodation" allows the coexistence of fishes and crayfish in the littoral zones of lakes. Interaction between this set of GLP and SDP, analogous to the fish/invertebrate example in lake limnetic zones, should provide inter-lake and within-lake seasonal shifts in predation pressures on the littoral prey communities that include fishes and invertebrate prey.

Interactive predation between fishes and large, benthic, invertebrates is common also in marine systems. Virnstein (1977) tested earlier suggestions about the importance of crab and fish predation in structuring infaunal abundances of prey organisms in Chesapeake Bay, Maryland, shallow-water subtidal communities. For two consecutive years he sampled infauna from caged and uncaged replicated plots; the cages contained blue crabs (*Callinectes sapidus*, Portunidae), sciaenid fishes (*Leiostomus xanthurus*), hogchokers (*Trinectes maculatus*, a flatfish, Soleidae), or none of these predators. In comparing caged and uncaged areas, he concluded that both crabs and sciaenid fish predation controlled prey density and species composition of the infaunal community, whereas hogchokers did not. When predators were artificially excluded, "opportunistic" prey species showed the greatest population increases, suggesting that these species were most subject to heavy predation pressures in natural situations. Although not considered by Virnstein, blue crabs are preyed upon by benthic-feeding fishes, and it is likely that fishes such as *Leiostomus* have a major controlling effect on crab densities, probably through predation on juvenile and intermoult individuals. In the more complex marine subtidal world there may be several predators whose interactions affect the infauna.

Another marine benthic invertebrate, the sea urchin, affects the species composition, relative abundances, and distribution of marine plants (Ogden, Brown, and Salesky 1973, Breen and Mann 1976 for tropical oceans; Paine and Vadas 1973 for temperate intertidal regions). Sea urchins also affect animal populations, including corals, by altering the substrate and by direct grazing, which may limit growth (Glynn et al. 1979). The latter study on the cidarid echinoid *Eucidaris tribuloides* indicated that the species grazed heavily on several types of live pocilloporid corals in the Galápagos Islands. Examination of the related species *E. thouarsii*, which is found in mainland populations of Ecuador and in reefs off Panama, indicated a relatively scarcer, smaller, more cryptic, and more nocturnal animal. Glynn et al. (1979) investigated the hypothesis that differential fish predation was responsible for the observed difference between the two species of echinoid. Adult *Eucidaris* (>10 cm test diameter) placed in the open off the Galápagos were not eaten; only subadult sizes were consumed by the labrid *Bodianus diplotaenia*. In contrast, when *Eucidaris* were placed in the open on Panamanian reefs, echinoids up to 3.58 cm were eaten within two or three hours by the balistid (triggerfish) *Sufflamen verres* and the tetradontid (puffer) *Arothron meleagris*. The authors conclude that fish predators may be more effective on mainland urchin populations, which accounts for the subsequent species differences. If fish predation is as important as they suspect, the species composition of subtidal plants and animals may be explained by the interaction between fishes and echinoids.

In northern oceans, the interaction of sea urchins with higher vertebrate predators may produce profound consequences on the littoral and sublittoral community. In the western Aleutian archipelago of Alaska, Estes and Palmisano (1974) compared the nearshore marine communities of Amchitka Island (Rat Island group) and Shemya Island (Near Island group), which have physical similarities and are in geographical proximity, yet present major floral and faunal differences in their lower intertidal, rock platform (bench) communities. Amchitka benches are almost covered by a mat of benthic marine brown algae (kelps) to 20 to 25 m, the dominant species being *Hedophyllum sessile* and *Laminaria longipes*. In addition, barnacles (*Balanus glandula* and *B. cariosus*), mussels (*Mytilus edulis*), and other important herbivorous invertebrates, such as sea urchins (*Strongylocentrotus*) and chitons (*Katharina tunicata*), are inconspicuous, small, and scarce. In contrast, Shemya Island is characterized by a lack of macrophytes below the intertidal zone. Barnacle and mussel densities are more than two orders of magnitude higher, and sea urchins almost completely carpet the sublittoral region in many areas.

Estes and Palmisano propose that the distribution of sea otters (*Enhydra lutris*) explains the dominance of sea urchins, which are in turn responsible for the difference in the macrophyte communities. Sea otters, which consume sea urchins as a major portion of their diet, are scarce in all but a few sites in western North America, a recent restriction from an original distribution that ranged from the northern Japanese archipelago through the Aleutians and as far south as Baja California. Amchitka is one of the few Aleutian Islands where abundant native otter populations have continued to exist since their decimation by Russian trappers during the eighteenth century. Amchitka otter densities have been estimated at 20 to 30 animals per square meter of habitat in the intertidal and sublittoral regions within the 60-m contour. Their control of urchin densities on Amchitka explains the differences in flora and fauna between this island and Shemya. The conclusions regarding the role of sea otters and their recent decimation by man have been recently supported by the data of Simenstad, Estes, and Kenyon (1978), who conclude from data including Aleut middens that man's recent exploitation has resulted in present distribution patterns. In this example it is the interaction between a dominant grazing invertebrate (*Strongylocentrotus*) and higher vertebrates, otters or man, that influence the littoral and sublittoral communities.

All of the previous studies indicate the ability of interactive predators to control community species composition. Studies of prey morphology and its relation to specific predators indicate that the observed community effects may be similar to those described for freshwater communities. The strong relationship between prey morphology and fish selectivity has been studied for fishes feeding on molluscs. Molluscivores such as carp (*Cyprinus carpio*) crush shells with their pharyngeal teeth, which have flattened molarlike surfaces. Ivlev's (1961) early experimental work indicated that carp feeding on the freshwater bivalve *Dreissencia polymorpha* selected the thinner-shelled individuals. Macan (1977) has commented on fishes feeding on the freshwater *Lymnae* snails. Stein, Kitchell, and Knežević (1975) found that fish electivity for carp feeding on several different types of freshwater molluscs in Skadar Lake, Yugoslavia, was highest for *Valvata piscinalis*, which had a readily crushable, thin shell, among other characteristics. All of these studies of freshwater molluscs indicate a strong relationship between molluscivorous fishes and prey morphology (reviewed by Vermeij and Covich 1978).

A series of studies of marine molluscs has shown a relationship remarkably similar to that for lake zooplankton exposed to two classes of predators. In marine systems molluscs are exposed to two important classes of predators, fishes and crabs. In addition, stomatopods (pistol shrimp) and lobsters may be locally important, although this has not been as thoroughly examined.

Crabs, which manipulate their prey much as freshwater invertebrate predators do, can be classified as either peelers or crushers. Unspecialized peeling crabs usually insert the fixed finger of the manus of one claw into the shell aperture and break the shell in an outward direction (Zipser and Vermeij 1978), which allows them to remove the mollusc's soft parts. Crushing crabs break the shells with their master claws and then remove the prey. Anti-crab morphologies include: low spires, which make peeling more difficult; reduced spines, which may be effective against crushing if they interfere with the ability of the crab to apply the master claw; and thickened shells, which are most effective against crushing by crabs. In addition, the size of the shell is critical, for the vulnerability of a prey for a given predator is zero (Vermeij 1976) above a certain point.

Although fishes have been recognized as mollusc predators for a long time (Vermeij 1978), the first experimental studies of the relationship of mollusc shell morphology as a defense against fish predation was only recently published (Palmer 1979). Fishes can attack molluscs as crushers, swallowers, or by feeding on the mollusc's feet. As with crab predators, the most general adaptation against fish predation is a thickened shell. Stout spines and reduced spires with internally thickened apices are considered by Palmer (1979) to be primarily adaptive to prevent predation by crushing by teleost fishes such as pufferfishes and rays. In contrast to the varied crushing modes of crabs, fish crushers use their jaws and can apply stress only to restricted areas of the mollusc shell. The only means of prey escape from those fishes that swallow their prey appears to be size, a conclusion drawn also by Stein and Kitchell (1975) from their studies with carp. The bladelike varices of molluscs are seen as a morphological adaptation for fishes that tear the feet off their prey. The blades help dislodged molluscs to land on the substrate with their feet down, thus protecting those parts of the body (Palmer 1977).

The classes of mollusc predators follow those outlined for freshwater communities. Each class of predator is selective in specific ways, and the prey adaptations relate specifically to the type of predator. Vermeij (1978) has been able to extend this analysis to questions of species diversity, latitudinal and interocean predation gradients, and population variations. He suggests that interoceanic morphological differences, as well as comparisons between ecologically similar warm versus cold water forms, can be related to differences in crab predation (Vermeij 1976, 1977, 1978). As with freshwater zooplankton communities, species morphologies can be predicted from a knowledge of which predators are present in a given locality. In addition, there is clearly a strong interaction between fishes and benthic invertebrates in these marine systems. Closer investigation should be able to identify prey morphologies that are adaptive specifically for situations in which both fishes and invertebrates coexist, as has been seen for the zooplankton species living in the presence of Class I and Class II predator types.

The anti-predator adaptations of freshwater molluscs are more conservative than those found in marine molluscs (Vermeij and Covich 1978). The spectacularly sculpted forms commonly found in the marine environment appear only in the oldest lakes, not because of limited calcium carbonate or other physical explanations, according to the authors, but because the diversity of predators necessary to favor a greater variety of successful anti-predator adaptations is simply not present in freshwater systems.

Although the relationship between predator pressures and prey morphologies has been recognized for centuries, as the previous discussion has indicated, this relationship can be, perhaps, more predictable than was previously imagined. Whenever species taxonomy is based on gross morphology, as in many invertebrate groups, community species composition can be predicted from a knowledge of interacting predators. The expectation that specific prey morphologies and behavioral patterns are associated with specific predator types can be extended as well to vertebrate assemblages such as fish communities (Zaret 1979). Competitive interactions are most predictive when applied to feeding morphology.

SOME NEW ANSWERS

If predation and competition are primarily responsible for determining the species composition of communities, nutrient levels should have their greatest effects on the numerical abundances of the given species. That is, an increase or decrease in nutrients, such as from eutrophication, should have greater quantitative than qualitative effects (Brooks 1969). Nutrient manipulations, natural or controlled, might produce population fluctuations but should lead to few species extinctions or additions.

Hall, Cooper, and Werner (1970) showed in experimental ponds that, although the addition of fish had strong effects on species composition, nutrient additions affected the numerical abundances of species. Fertilization (Losos and Hetesa 1973) likewise produced increases in zooplankton densities but not changes in species composition. Hrbáček (1977) presents some evidence that lower concentrations of food produce larger individual zooplankton of the same species. Other data also support this conclusion. Studies by East European scientists working on carp ponds were the first to show that changes at higher trophic levels (specifically fish) produced not only alterations in zooplankton species composition but also in the composition of primary producers, the algae (Hrbáček et al. 1961; Hrbáček 1962). This has been supported by more recent work with fishes in small laboratory experiments (Hurlbert, Zedler, and Fairbanks 1972*b*), artificial ponds (Hall, Cooper, and Werner 1970; Losos and Hetesa 1973), and large lake fish introductions (Zaret and Paine 1973). Other work on grazing effects by zooplankton illustrates how the co-evolution of zooplankton and phytoplankton populations leads to certain predictable patterns of phytoplanktonic dynamics (Porter 1976), which suggests that predictions for zooplankton prey would be made also for lower trophic levels. For instance, because some zooplankton species have a more rapid, efficient, or selective means of grazing, predictions for algal composition based on levels of gape-limited predation and the effects of lake nutrient levels are possible.

The theory of interactive predators provides plausible answers for other questions as well. From his considerable work on the genetics of daphnids, Hebert (1974) asks why, if given the great variation in genetic polymorphism among populations, even within the same lake, the morphology of the species should remain constant. This appears to be true even when comparing populations separated by large geographical distances. One answer is that the morphology of the genus *Daphnia* is determined by specific predators, which are likely to be present in many lakes of a given region. Thus the morphology would be correlated with predator geographical distribution. Genetic polymorphism, on the other hand, may include feeding, physiological tolerances, or aspects of competition, which are more likely to vary between lakes.

Littoral cladocerans usually number anywhere from two to five times more than the limnetic cladocerans in a given lake. Given the basically homogeneous nature of lake limnetic zones, prey species can avoid predation in only a very limited number of ways (see Zaret 1975), which results in a limited number of morphologies. Whereas the limnetic regions are without the prey protection afforded by physical habitat, the littoral zone has considerable physical structure in sediment and aquatic macrophytes, which provide potential subhabitats or prey refugia. For instance, large forms such as crayfishes or freshwater crabs can escape under rocks. Thus an increased diversity of the physical habitat should allow a greater number of prey niches and a higher number of community species, thus explaining why "environmental heterogeneity" is associated with a more diverse species assemblage (e.g., Smith 1972).

The concept of anti-predator devices provides a hypothesis for daphnid helmets, an adaptation that was formerly considered to be due strictly to physical factors (see Dodson's references 1974*b*). This may not be true for all cases, but it is certainly documented for several situations.

One potential long-range use of these models is the identification of the predation state of a lake by identifying the prey species in the community. If one finds the fairy shrimp, *Branchinecta*, or large numbers of *Daphnia pulex*, one can be certain the system is without gape-limited predators and is in State 0. If instead one finds helmeted daphnids, one can be confident that both gape-limited predators and size-dependent invertebrate predators are in the lake. If *Bosmina* is common in the limnetic zone, one can assume that gape-limited predation is intense and that size-dependent predators are in low abundance. If one has some basic biological knowledge of the region or the lake type, other intermediate states can be identified. The best way to test the precision and accuracy of the model, however, is through direct manipulation, that is, to shift the system from one state to another. For instance, if we add another planktivorous fish species or artificially increase the numbers of size-dependent predators or vary nutrient levels, shifts in species composition are predicted. Some tests of this kind have been undertaken with results compatible with the model. Hrbáček et al, (1961) studied East European ponds before and after the introduction of carp; Hall, Cooper, and Werner (1970) and Hurlbert et al. (1972*b*) used experimental ponds in which to vary nutrients or introduce fish. In each group, a shift in zooplankton community composition occurred, along with increased gape-limited predation. Large-scale manipulations might relate to sensible ecosystem management of natural systems. Shapiro, Lamarra, and Lynch (1975) call such a procedure biomanipulation and cogently argue that it not be limited only to nutrient additions, as has been true usually, but that it should employ biotic information as well. Thus, this theory could be used as a probe for initial explorations into the extent of biotic perturbations of natural systems in ways that might relate to the practical problems, such as lake eutrophication, that presently affect our society.

The relationship between predation pressures and the productivity of a system is not well known. Although Steele (1974) included predation by fishes in his model of marine ecosystems, he nevertheless considers it relatively unimportant because calculations of energetics do not indicate fishes as a major component of the total biomass of the oceans. In contrast, Vinogradov (1962) previously pointed out that predation is the major mechanism of energy transfer in the deep-sea zooplankton. These two contrasting points of view illustrate the different ways of considering a system. In the former, fishes become relatively unimportant from a consideration of biomass or energetics. In the latter approach, fishes, as the mechanism or determinant of community species composition, assume a major role. Although Steele is undoubtedly correct in his calculations, it would be folly to ignore the effects of higher trophic interactions as the mechanism of community structure, especially if different communities demonstrated significantly different total productivities, which seems possible from freshwater studies. Both Hrbáček (1962) and Brooks (1969) point out that considerations of biomass alone are insufficient to analyze the dynamics of aquatic ecosystems. As an obvious example, man's total biomass on earth is relatively insignificant, yet clearly his actions could alter or destroy the entire system.

Lake sediments contain fossil remains that are used to reconstruct past abundances of zooplankton (Goulden 1971); pollen deposits indicate the former extent of forests (Davis 1969); earlier phytoplankton populations can be derived from diatom remains (Bradbury 1971), and previous nutrient input can be based on changes in ions (Cowgill et al. 1966). We can now attempt this for the entire biotic community (e.g., see Brooks 1969; Kerfoot 1974*b*). By considering the prey species and their morphology in the sediment, predictions can be made about what kind of predation pressures were present over time.

THE CONCEPT OF NICHE

Throughout this book the term "niche" has been used in Elton's (1927) functional sense, that is, as the position or role of a species within the community. A return to this functional-niche concept has found recent favor with others as well (e.g., Williams 1975). This usage may be more meaningful for a treatment of species community structure than the n-dimensional hyperspace proposed in Hutchinson's earlier, influential paper (1957), in part because of the blurring of this latter concept in recent literature. This confusion has been recognized, and a recent paper proposes to redefine niche by using Hutchinson's graphical approach but restricting its meaning to that first suggested by Elton (Whittaker, Levin, and Root 1973).

As defined by Hutchinson (1957), the fundamental niche boundary is determined by the limiting states of all variables, including physical and biological ones, that will permit the species to exist indefinitely. The realized niche is that portion of the fundamental niche within which the species is constrained by competitors. As insightfully pointed out by Connell (1975), this seminal definition placed great emphasis on the effects of competition but basically ignored the equally important biotic factor of predation. Rather than redefine the niche concept to include predation, many workers have relegated predation to one of the niche parameters; that is, the ability to avoid predation is considered an advantage in competition between species (see Hutchinson 1975). The weakness of this approach has been pointed out by many, because if predation keeps prey population levels low enough so that resources are abundant relative to their numbers, competition between them will not occur (see Birch's commonly accepted definition of competition, 1957). That predation reduces species competition has been documented in the field (Paine 1966; Zaret 1972*a*), thus providing support for the earlier theoretical suggestion of this possibility (Slobodkin 1961). To consider community structure without understanding the effects of predation is a serious conceptual error (see Connell 1975 for a discussion of this problem).

The question becomes most basic when the cause of species extinction is considered. A competition argument would state that extinction occurs when one species is excluded from a required resource by another through competitive exclusion. A predation approach would argue that predators can directly cause species extinction. This argument about the causes of extinction has its roots in the earliest evolutionary literature, in the contrasting opinions of Darwin and Wallace about the nature of natural selection (see preface). In fishless lakes, for instance, large-sized herbivores such as *Polyartemia* can persist indefinitely. Once planktivorous *Salvelinus alpinus* are added, however, the changing conditions result in the elimination of this species as well as its niche (the latter is indicated by the fact that other species of equally large-sized herbivores do not enter the lake). Instead, we see the appearance of a medium-sized herbivore, *Daphnia galeate*, not found previously in the lake. This indicates the presence of a new niche in the system. Numerous other examples have been discussed previously.

If the niche is functionally defined as the selective forces defining it change, including the physical environment, predation, and competition, the niche disappears as well as the species associated with it. Other niches may then become available, and other species, new or from within the community, will fill and compete for the remaining resources.

As a slight modification of Hutchinson's concept of the niche I propose the following. *The fundamental niche boundary is determined by the limiting states of all physical variables that will permit the species to exist indefinitely. The realized niche is that portion of the fundamental niche within which the species is constrained by predators and competitors.* Using this definition, we say that for these freshwater zooplankton communities, predation defines that aspect of the realized niche that we call gross morphology; competition determines that aspect of the realized niche that we call fine morphology, especially relating to feeding morphology. This probably explains why scientists examining a guild of species (species feeding on a similar resource such as herbivorous copepods) often conclude that competition is the dominant biotic factor, whereas scientists examining several trophic levels simultaneously emphasize predation. Both conclusions are correct but are limited to the sphere of examination.

Whether predation or competition is more important is a rather meaningless debate. Of more concern is, "Under what condition is predation or competition the dominant biotic factor?" which implies the importance of both processes and seeks to determine when and why one is dominant over the other. Connell (1975) concludes that competition is dominant in physically harsh environments, where many predators are unable to live, whereas predation is the dominant process in benign environments. In a study based exclusively on marine intertidal areas, Menge and Sutherland (1976) conclude that predation determines the number of species in a community of lower trophic levels, whereas competition regulates community species number at higher trophic levels. These two papers present contrasting conclusions to each other and also to those proposed in this book. If nothing else, I hope this book has served to put to final rest the debate over which—predation or competition—is the more important process.

Appendix: Further Considerations

CHAPTER 2

The laboratory feeding experiments of Werner and Hall (1974) suggest that an energetic basis for fish selection may exist and that fish choose prey items according to a maximization of energy per unit time. Although energetics may be the ultimate reason for a predator searching technique, it is not an explanation for the mechanism of prey consumption, which must be based on sensory information such as prey visibility, motion, or palatability. The most important conclusion of the Werner and Hall study is that fish select the largest prey at a greater rate than expected because the largest prey are the most easily detected. If this is true, given a mass of prey items and the possibility that somewhere a still larger prey exists, how do the fish learn to take the largest prey? This implies an ability to census the prey environment before feeding, and there is no current information available that suggests fishes have this ability.

Vinyard and O'Brien (1975) found that bluegills in aquaria exhibit a dorsal-light response that correlates with prey body-size. The "dorsal tilt" was first described, according to Curio (1976), by Holst (1948) in a study of *Pterophyllum eimecki* (Cichlidae), the popular angelfish of aquarists. Holst found a correlation of reaction angle and hunger level. Vinyard and O'Brien (1975) found a correlation with prey body-size. They also found that bluegills show no difference in dorsal tilt when presented with two prey where one has a greatly increased amount of pigment or where one prey has increased motion. Others have shown, of course, that both these characteristics result in increased prey selectivity by fish. It is not clear why the dorsal-light response occurs with different-sized prey but not with prey of the same size that differ in visibility or motion. The interpretation would be easier if the adaptive significance of the fish behavior were known.

The fish eye is relatively inefficient at distinguishing objects, and most fishes are considered to be myopic. Verrier (1948) suggests that the teleost eye is basically suited for detection of movement and is defective in the perception of form. Protasov (1968) indicates that the ability of fish to detect motion correlates with light levels. Suffern (1973) has shown that both the fish feeding rate and the degree of fish selectivity on prey decreases when available light is lowered. One wonders which prey characteristic would become least detectable with decreasing light conditions, prey visibility or prey motion?

If planktivores feeding on zooplankton do not filter-feed, why is there a correlation of gill-raker *number* with prey size (Kliewer 1970) or with the habit of plankton feeding (Nikolsky 1963)? To date, no one has studied the mechanism of gill-raker use in planktivores. It is possible that gill rakers function by trapping particle-selected prey until they can be swallowed by the fish.

In all environments food resources are not distributed homogeneously in space but assume some form of clumped or hyper-dispersed distribution. Zooplankton in lakes, for instance, are located in patches over horizontal as well as vertical spatial gradients. A number of recent theoretical papers have examined the origin of plankton patches (Levin and Segel 1976) and the consequences for predators of patchy prey distribution (Oaten 1977). On the empirical side, Hunter and Thomas (1974) made the observation that when larval anchovies of *Engraulis mordax* encounter a prey patch, they not only remain in the patch to feed but also increase their feeding rate. Lasker and Zweifel (1978) have presented further data supporting this. Some very exciting investigative possibilities exist once the theoretical information about plankton patches and its consequences for planktivore feeding patterns is developed.

The function of the cladoceran eye is still unresolved. Although zooplankton do not require a compound eye for vertical migration (Harris and Mason 1956), it might nevertheless be necessary in the location of phytoplankton. When it is discovered that only the eye size increases and not the number of ommatidia (Wawrik 1966), one wonders what the advantage of a larger eye is. One possible clue is that predatory species possess more ommatidia (several hundreds) than filter feeders (in the tens), as found by Wesenberg-Lund (1939).

Recent work on the developmental process of the *Daphnia* eye (Macagno 1978) may help to provide answers to this question.

CHAPTER 3

Fish planktivores always feed in a school rather than as solitary predators when feeding on zooplankton. This is true not only for different planktivorous species; given two closely related species where one feeds on benthic prey and the other feeds on plankton, we find that the former hunts alone, whereas the latter always feeds in a school. (See Fryer and Iles 1972, which deals with the African Great Lakes and species flocks of cichlids.) Even the same individual that feeds at certain times of the year on plankton and later changes to benthic organisms will switch from schooling to individual feeding. One explanation is that schooling prey receive better protection from predators than solitary prey (e.g., Neill and Cullen 1974). This apparently includes *Daphnia*, whose nonrandom distribution is often in patches, although it is not usually considered a schooling animal. Experimental work by Milinski and Curio (1975) on the advantages enjoyed by schooling daphnids suggests an alternative explanation for zooplankton patches which could be considered either a consequence or an adaptive response to predation by visually hunting planktivores.

It is unclear what kinds of information are necessary to evaluate the possible effects of planktivores having greatly differing rates of prey consumption. The maximum rate of feeding for *Ambystoma* in the Colorado alpine lake studies was approximately 100 prey items per day, and often the figure was much lower. For fish, however, the figure may often approach and sometimes exceed two orders of magnitude above this (i.e., 10,000 items per day for some coregonids). For insect planktivores the value may be one order of magnitude below that for *Ambystoma*.

Learning in fish is another very important yet poorly understood and rarely studied phenomenon. Ivlev (1961) has shown that one can actually train a captive fish to feed on food items it would normally abhor in its native habitat. Nilsson (1963) showed that by feeding trout the least preferred of two artificial food types, the fish would continue to feed on this least preferred type for a short period of time, even when the preferred type was a higher density. This learning ability means that the consistent use of the same individual fish for repeated feeding experiments in the laboratory would allow the predator to change its behavior over time as a result of its laboratory experiences. One can easily observe this by feeding fish in an aquarium by hand over several weeks. Soon the fish will respond to the approach of a human being by coming to that side of the tank in anticipation of food. In fact, one can eventually induce fish to commence feeding movements by merely tapping on the tank or disturbing the water, even if there is no food present, in the same way that Pavlov was able to make dogs salivate by ringing a bell. The conclusion, that fish feeding experiments must be very carefully controlled, indicates that to achieve any kind of meaningful results many individual fish must be used as experimental subjects.

Competition theory predicts that two species feeding on similar-sized food items will either alter their normal preferred diets, diverging from each other, or one will exclude the other. In the Colorado alpine ponds, Type B communities are dominated by one predator, *Diaptomus*. The arrival of *Ambystoma* changes the assemblage to Type A, and the resultant community has two different predator consumers, *Chaoborus* newly added, each feeding on different-sized prey items, not as a result of competitive divergence but from shifts in prey-size distributions as a result of *Ambystoma* feeding preferences. Current theory is insufficient to distinguish whether feeding by two consumers on different-sized fractions of prey is caused by competitive displacement or complementary predation.

CHAPTER 4

Strickler and Twombly (1975) have speculated that capture success is not strictly a size-related phenomenon because copepod predators are keying in on sensory information that relates to both size and velocity. A large, slow prey may present the same information as a smaller, faster one. It is unknown how these copepods could use sensory information to make this distinction.

Any factor that decreases a predator's feeding efficiency should, in theory, act to stabilize a predator-prey relationship. Copepods are capable of a rheotactic response and initiate escape movements upon detecting violent water disturbances, including those from large conspecifics. One consequence of increases in their population densities could be an increase in the frequency of escape movements, which would reduce their feeding times. This might serve as a density-dependent feedback loop, of considerable importance for foraging models involving copepods. There is a strong possibility that calanoid and cyclopoid populations would show different results.

Pioneering work on invertebrate predation by Holling (1966), modeling the "functional response" of mantids (i.e., the shape of the curve describing predation rate), suggests that invertebrate predators are not able to cause the extinction of prey species. As prey densities become lower and lower, the invertebrate has such trouble locating prey that its numbers diminish before it can remove the last prey. This is supported by earlier studies on mites by Huffaker (1958). Size-dependent predators may decrease prey abundancies, but some other cause, such as an adverse environmental fluctuation or the effect of competition from other prey species, may be required for the final death blow. If true, this would be a major departure from the abilities of gape-limited predators, which are capable of driving prey species to extinction.

A recent model for predatory copepods (Gerritsen and Strickler 1977), based on the previous encounter models of Holling, examines the detection field of these nonvisual predators. The shape of the encounter field no doubt differs for visual and nonvisual predators, but the consequences of these differences, in terms of qualitative and quantitative effects on the prey populations, also are unknown.

These are some of the many fascinating problems that remain to be explored and elucidated. A further, perhaps obvious, question comes to mind: if predatory copepods use only tactile cues for prey capture, what function is served by their compound eyes?

The world of the predatory copepod is based on mechanical rather than visual stimuli. When a prey such as *Bosmina* ceases motion, it is equivalent to stopping the flow of information or turning out the lights for a visual predator. Other, nonvisual information signals, such as smell (pheromones), which are so important for other animal groups, may provide supplementary information. Few other responses are available to prey having to escape from tactile predators such as copepods.

CHAPTER 5

If vertical migration is such an effective escape response, why don't all prey exhibit this behavior? An interesting answer is found in recent studies by Begg (1976) of Lake Kariba in Africa. The sardine *Limnothrissa miodon* was the dominant planktivore, feeding most heavily on *Bosmina longiros-*

tris. Another plankter, *Mesocyclops leuckastii*, was also abundant, but it was not eaten by *Limnothrissa*. If one examines the vertical migration pattern of these two crustaceans, one finds that *Mesocyclops* descended rapidly from surface waters at dawn, independent of the position of the thermocline, as long as these deeper, bottom waters contained 1.5 ppm of oxygen, a rather low value. *Bosmina*, however, under laboratory conditions, was not able to tolerate such low oxygen levels and presumably could not migrate as deeply as *Mesocyclops*, thus leaving it more susceptible to predation during the day by *Limnothrissa*.

It is important to remember that morphological characteristics may be correlated with predation pressures rather than with one specific type of predator. For instance, Arcifa (1976) has found that in populations of the helmeted daphnid *Daphnia gessneri* in Brazil there is a significant difference in eye diameter of 12.5 percent between the two morphs. In this example, however, it is the high-helmeted morphs which have the larger eye, contrary to results for north temperate daphnids. This exception does not mean that reduced eye diameter is unrelated to predation pressures, but that the visual predators in Brazilian reservoirs may be feeding more selectively on the smaller helmeted forms. A test of this suggestion will have to await ongoing studies of the role of visual predation in these Brazilian zooplankton communities.

The theory for visually feeding fishes and their influence on prey community structure can be extended to nonlacustrine freshwater communities. Allan (1978) made observations on patterns of insect drift in Colorado streams, noting that there was a correlation between the nocturnal emergence patterns of the abundant mayfly (*Baetis bicaudatus*) and its presumed susceptibility to fish predation by brook trout (*Salvelinus fontinalis*). Allan's results suggest that the larger the emergent individual, the more likely it will emerge during the evening when visually feeding fish have a reduced efficiency. The study bears a remarkable relationship to the finding for vertical migration patterns of crustacean prey in lakes with visually feeding fishes (Zaret and Suffern 1976). Allan is probably the first investigator who has attempted to bring to studies of stream invertebrates the considerable theory of biotic interaction developed from lake systems. The applicability of these principles to stream systems is potentially one of the most fertile future areas of research in freshwater ecology.

Recent work (Hairston 1976, 1977) has indicated that carotenoid pigmentation in copepods protects against visible light waves, which otherwise can cause death in unpigmented animals. This damage from visible light is due to the absorption of light by sensitizer molecules, which transfer energy to oxygen. The higher energy oxygen may then bind to other intracellular compounds, resulting in impaired functioning or possibly the production of toxic compounds. The carotenoid pigments act as quenchers for the excited oxygen. Hairston (1977) also found, however, that strong pigmentation increased the likelihood that predatory salamanders would feed on the copepods. The implications of this in relation to copepod distributions among lakes and in vertical distributions present an interesting problem.

Unpalatability, one of the most commonly successful anti-predator adaptations found throughout terrestrial invertebrate taxa (Edmunds 1974), is conspicuously rare in freshwater systems. This general absence is surprising since the rapid learning ability of fishes makes them potentially vulnerable to this kind of prey evolutionary strategy. The only non-piscine examples of aquatic prey include the aforementioned *Bufo* tadpoles (see p. 46 above) and some water mites (Riessen pers comm.), both of which appear toxic to fishes.

There are several possible explanations for the absence of unpalatable zooplankton in freshwater systems. First, perhaps the chemicals necessary to produce toxins in zooplankton are either unpredictable or unavailable. In terrestrial environments, for example, many invertebrates develop toxicity by ingesting, with impunity, the noxious chemical compounds concentrated in plants. These compounds are stored in the prey where they are accessible to potential predators (e.g., monarch butterflies, *Danaus*, feed on milkweeds, *Esclepias*, and concentrate the poisonous toxins in their wing scales; attacking bird predators learn to avoid them). In fresh water, some blue-green algae have toxic chemicals associated with them, but perhaps they are not available for storage by zooplankton.

Another possibility relates to the theory of aposematism (warning coloration) in toxic prey. Even the most noxious prey gains no advantage when mistakenly killed and eaten by a naive predator. For the evolution of toxicity in zooplankton, the predator must learn to avoid the noxious prey, and this is facilitated by conspicuous, warning prey color patterns. If piscine planktivores feed intensively at dusk, however, when their visual acuity and especially color sensitivity are low, it may not be possible to develop effective aposematic colors.

Finally, the relationship between fish size and prey size may also provide an explanation for the absence of toxic zooplankton in lakes. In theory, for an individual zooplankton to benefit from being unpalatable, the predator must be able to attack and reject the prey unharmed. This seems unlikely in the case of predators which swallow their prey whole or, as in the case of most fishes, are more than ten times the size of the prey particles. This suggests that unpalatability is more likely to be present only with relatively larger prey; indeed, marine open-water systems do have examples of toxic zooplankton, including the Ctenophora.

Predators consume prey species directly, but their indirect influence is even more pervasive. The zooplankton individual that migrates to avoid predators is expending energy that could otherwise go into egg development. The animal that develops a helmet to reduce invertebrate predation or has reduced eye size to avoid fish predation is sacrificing a number of potential offspring or its ability, perhaps, to find food. The animal that avoids the best feeding areas or visits them only at night to escape predation is restricting its ability to accumulate energy and is also sacrificing potential offspring. The cost of genetic, morphological, or behavioral "predator insurance"—the indirect effects of predation—may put constraints on population distributions, densities, and reproductive abilities that are more significant than the small percentage of individuals actually removed by the predators.

CHAPTER 6

Increasing fish predation is correlated with cladoceran species extinction, as suggested in a recent study (Gophen, 1979). However, predation may not be the sole factor responsible. An interesting hypothesis to partially explain the extinction by fish of such daphnids as *D. pulex* has been proposed by Heisey and Porter (1977), who show that *D. magna*, the pond species, has a higher filtering and respiration efficiency in laboratory-controlled oxygen concentrations than does *D. pulex*, which has evolved to live in lakes. The authors suggest that the respiratory pigment hemoglobin, abundant in *D. magna* and less so in *D. pulex*, is responsible for these differences, and that fish predation in lakes prevents its development in the latter species because an increase of dark pigment would increase its conspicuousness. They hypothesize that when fish are present, *D. pulex*, unable to escape by migrating deeply into poorly oxygenated waters as a consequence of its lack of well-developed hemoglobin, can be forced to extinction. This example illustrates the interaction of biotic and physical features in lake community dynamics, and may also explain why vertical migration is not always successful for prey. Otherwise we might expect to see *D. pulex* present under conditions of even intense planktivore pressure. Empirical evidence shows that fish can cause the extinction of this species in spite of its vertical migration patterns. Whether this is due to physical limitations as found by Heisey and Porter or biotic interactions between the predator and prey populations may vary according to the lake.

For the purposes of the model construction and piscine planktivore flow charts (fig. 15), I have assumed that no higher level predators are present. If we added a piscivore capable of consuming any lake planktivore, consistency would dictate that the particular predatory fish would determine to a great extent the prey fishes that would persist in the system. There would be, at least initially, behavioral accommodation, such as the maintenance of tight schooling patterns in the planktivores, restrictions to habitats less vulnerable to predation, and other evolutionary changes similar to those occurring in the zooplankton (see Eggers 1978). These might be used to explain seasonal changes or lake succession.

Several recent studies have presented confirmation of the theoretical predictions presented in this chapter. In the lakes in the Toolik region of Alaska O'Brien and Schmidt (1979) found the discussed correlations between *Bosmina* size, development of exoskeleton, and eye reduction with vertebrate and invertebrate predators: where fish predation is intense, the predatory copepod *Heterocope septentionalis* is absent and the crustaceans have the smallest size and smallest eye diameters; where *Heterocope* and fishes are present, the *Bosmina* have the most developed mucrones and antennules. The authors were able to demonstrate from experimental studies that the *Bosmina* from lakes with predatory copepods were better able to resist the attacks of *Heterocope* than *Bosmina*, even of twice the size, from lakes without this invertebrate predator.

Of the size-dependent predators associated with temporary ponds, one of the most important groups is the odonates such as those of the family Aeschnidae, which attain sizes of up to several centimeters and more (Kime 1974). Some preliminary experiments with *D. magna* and *D. pulex* and temporary-pond odonate nymphs seemed to indicate that, whereas all *D. pulex* could be eaten, a certain percentage of the largest *D. magna* might escape (J. McKenney, unpublished data), suggesting that the unusually large sizes attained by *D. magna* are adaptive for the large pond predators such as odonates. It appears that *D. magna* and *D. similis* have evolved specifically for temporary ponds and thus possess characteristics adaptive for that existence; in the process, they have sacrificed a certain amount of competitive ability, which makes them unable to establish populations in lakes when *D. pulex* is present. (See chapter 8 for further discussion of competitive abilities.)

When fish predation first increases in a lake, the largest cladoceran species may show dramatic decreases in the population mean size of mature instars, and often mature instar sizes are at their minimum. Because early maturation results in increased reproductive rates, one might expect cladocerans to mature at these smaller sizes in the absence of fish predation. There must be some evolutionary disadvantage to maturing at these smaller sizes.

Taylor (in press) has recently used Leslie matrix models for predicting life-history responses of zooplankton populations undergoing size-selective predation. She concludes that prey body-size is a better predictor of predation pressures than predation rate, duration of prey vulnerable period, or estimate of prey reproductive value. She also predicts that predation on adults should have a greater effect on cladoceran than on copepod population levels because of the particular life-history characteristics. It would be interesting to know what aspects of life history would most affect population densities given predation on adults only, predation on young only, and predation on only the oldest and youngest, which is the normal pattern found in higher mammals (see Curio 1976).

The many species of the cosmopolitan genus *Ceriodaphnia* (Burgis 1967) possess small protuberances on the exoskeleton called fornix spines. Stromenger-Kletowska (1960) reported cyclomorphosis in these spines for pond species of *C. reticulata, C. megops, C. pulchella, C. quadrangula, C. affinis, C, setosa,* and *C. rotunda*. It is likely that these spines are defenses against small invertebrate predators such as the rotifer *Asplanchna*. For this predator, the small spines of its prey can make a significant difference in mortality rates.

Another cladoceran whose mucrones (Stromenger-Kletowska 1960) and tailspines (Dodson 1974*b*) undergo cyclomorphosis is the genus *Scapholebris*. This littoral cladoceran may be developing anti-predator morphologies in response to cyclopoid copepod predation.

CHAPTER 7

In the examination of the northern Swedish lakes, the species *E. coregoni* appeared to be common in each type of lake. This may be an artifact of sampling, in that Bosminidae are known to move from their normal littoral habitat to the open-water zones only when the other limnetic cladocerans are removed by predators (e.g., Brooks 1969; Noble 1975).

The earliest approach to the examination of zooplankton community composition was the development of correlations between certain species and the trophic (nutrient) or chemical conditions present. This descriptive correlative technique still has uses today (e.g., Patalas 1975; Pejler 1975). The major criticism of this approach is that causality does not necessarily follow from correlation. For example, Stenson (1974) affirmed that the absence or presence of *Holopedium gibberum* correlates with lake pH (as found by early studies) because fish populations were also correlated with these water types. When, more recently, fish composition changed, *Holopedium* distribution no longer correlated with these physical factors. The previous assumptions of causality with pH collapsed because the reasons for the cladoceran distribution had not been accurately identified. Alternative explanations had not been originally developed.

The limnetic regions of Lake Titicaca, an immense, high mountain lake (>4,000 m) on the borders of Bolivia and Peru in South America, contain very few cladocerans in spite of the fact that planktivorous fish are not known to occur in most of these areas. Instead, the limnetic zooplankton is dominated by calanoid copepods, with cladocerans such as *Daphnia* and *Diaphanosoma* found in littoral areas where fish, including some planktivores, are more abundant. It would be interesting to speculate on the reasons for the absence of large cladoceran populations from the lake limnetic zones. We note also the absence of cladocerans from the world's oceans, even though cladocerans occur in such saline waters as inland salt ponds and estuaries. In the oceans, copepods dominate numerically and form important food items of larval fish diets such as clupeids (Marshall and Orr 1964).

CHAPTER 8

Leptodora is one of the few large invertebrate predators that can persist in the presence of moderately heavy fish predation pressures. Interestingly, *Leptodora* is apparently the only known cladoceran that does not contain hemoglobin (Fox 1948) or any carotenoid pigments (N. G. Hairston, Jr., pers. comm.). The resulting extreme transparency may be the main reason that it can develop dense populations even when fish are present. One consequence of this hemoglobin loss, however, is the very strong oxygen dependence of *Leptodora* (Moshiri, Cummins, and Costa 1969). It cannot live in low oxygen waters. How this attribute affects its competitive ability with other SPDs requires further study.

The major conclusion of studies comparing the competitive abilities of the two morphs of *Bosmina longirostris* (Kerfoot 1977*b*) was that the long-featured morph gave up something when it opted for better protection against copepod predators and that this cost of defense resulted in its relatively inferior competitive ability. There is another way, however, to interpret these data. Suppose that the poorer competitor of the *Bosmina* or *Ceriodaphnia cornuta* morph not only has a superior anti-predator defense but also is adapted to a different environmental condition. For instance, the short-featured *Bosmina* morph was a superior competitor at all experimental stations except in late summer, the period of the year when algal food resources are reduced and probably limiting (Kerfoot, 1977*b*). Perhaps the long-featured *Bosmina* morph is not only better at predator defense but is adapted also to that time of the year when predation is heavy and food is reduced—the mid- to late-summer period. What appears to be the inferior competitor during periods of food abundance may be the superior competitor during periods when food is scarce. It would be interesting to run tests of competition at several food levels to examine this possibility.

Lane (1975) investigated the importance of competitive interactions by testing the hypothesis, first proposed by Dumont (1972), that zooplankton vertical migration patterns could be a means of avoiding spatial and feeding overlaps with competitors. Although her overlap values suggested that this was not an effective means of avoiding competition, Lane may have been focusing on species that were not competitors, thus confounding her conclusions. For instance, the strong negative correlation of *Bosmina* and *Daphnia* undoubtedly is related to predation effects, as shown by the work of Noble (1975), rather than to the likelihood that they were avoiding competition. Perhaps vertical migration as a means of avoiding competition can be tested by incorporating the recent finding of Haney and Hall (1975) that zooplankton do not feed continuously over the day but appear to have specific diel feeding activity peaks.

Competition among zooplankton would be very different if small cladoceran species are more efficient on smaller particle sizes than larger species, as suggested from Neill's 1975 laboratory work and more recently by Frost.

Several investigators have suggested that large cladocerans in fishless lakes are replaced by small ones in late summer because of food limitations

(Clark and Carter 1974; Kwik and Carter 1975). Alternatively, competition may be aiding a process initiated by planktivore predation. Field experiments must be devised to distinguish between the importance of these two processes in effecting this change. It is possible that food limitations upon herbivores may account for this change in species composition.

The generalization has been made that cladocerans are favored over copepods in lakes with eutrophic conditions (McNaught 1975; Hrbáček 1977), although this appears to result more in dominance by some cladoceran species than changes in the species composition itself (Korínek 1972). One explanation for this change may be that food concentrations in eutrophic conditions may be adequate for cladocerans but not for copepods (Weglenska 1971). Gliwicz (1977) and others have expanded on this by suggesting that the type of blue-green algae normally associated with eutrophic conditions may make feeding more difficult for the larger cladocerans especially. This may provide an alternate explanation for the decrease in occurrence of large-sized zooplankton in eutrophic lakes, which has usually been attributed to associated fish predation.

CHAPTER 9

By developing predator electivity curves, rather than relying on individual predator characteristics, the effect of predation can be used to develop predictions of resultant community species composition. Models are used to develop testable predictions, to identify gaps in our knowledge, and to further our understanding of natural systems. To construct a predictive model one must (1) identify the dominant predators; (2) develop classes of predator electivity curves; and (3) define the range of prey evolutionary responses. As with any biological model, one does not expect to be able to mimic precisely the real world.

I have used freshwater communities to illustrate the use of this method, but the approach can be used for any system, freshwater, marine, or terrestrial. For instance, to construct a comparable model for the open ocean, one needs to identify the number of different selective pressures associated with predation. One might consider fishes, starting first with the planktivores. Because the size range of marine zooplankton is much broader than that associated with freshwater systems, we might consider all gape-limited predators as exhibiting a specific size range over which they could feed. Thus, instead of seeing a single figure 1 curve in which the predator is capable of taking any prey item within its habitat, we would construct a family of figure 1 curves, each covering a particular size range with each range depending on the body-size and associated gape of the fish. For size-dependent predators, again we would expect a family of figure 17 bell-shaped curves, each covering a distinct size range. There might also be predator specialists, such as the Ctenophora (comb jellies), whose prey-capturing nets make them the spiders of the open ocean, feeding on zooplankton prey (Reeve, Walter, and Ikeda 1978). This would introduce several distinctive selective forces associated with predation and a series of specific size-range curves within each of these types. Assembling the model would be considerably more complex than for a freshwater system, but using the same general theory, one could at least examine subsets of the oceans to apply it with the same efficacy as one can for lakes. Historically, however, studies of oceanic open-water communities have relied primarily on understanding physical process first and biological species interactions last. The field of biological interaction among predators is only beginning to be explored (e.g., Landry 1978).

If the success of a planktivore is intimately tied to its food resources, one cannot be confident about the eventual establishment of a fish without knowing something about the zooplankton community. This presents a problem in which considerations of resource substitution (e.g., Covich 1972), of switching in predators, and of similar phenomena are important.

Certain animal groups dominant in marine systems are completely absent from fresh waters. These include entire phyla, such as echinoderms (including starfish and sand dollars) and tunicates (sea squirts), among others. The absence of these phyla might be explained by considering water balance relations under salt- versus fresh-water conditions. But why are barnacles, a crustacean, absent from lakes? Why is there a shortage of sessile, filter-feeding animals, the dominants of the marine world? Colonial bryazoans do assume this role, but freshwater sponges are relatively uncommon in large lakes, with the spectacular exception being Lake Baikal, where there exist several sponge families, including the endemic Lubomirskiidae.

Bibliography

Adolph, E. F. 1931. Body size as a factor in the metamorphosis of tadpoles. *Biol. Bull.*, 61:376–86.

Allan, J. D. 1973. Competition and the relative abundance of two cladocerans. *Ecology* 54:484–98.

———. 1974. Balancing predation and competition in cladocerans. *Ecology* 55:622–29.

———. 1978. Trout predation and the size composition of stream drift. *Limnol. and Oceanog.* 23:1231–37.

Ambler, J. W., and Frost, B. W. 1974. The feeding behavior of a predatory planktonic copepod, *Tortanus discaudatus. Limnol. and Oceanog.* 19:446–51.

Anderson, R. S. 1967. Diaptomid copepods from two mountain ponds in Alberta. *Can. J. Zool.* 45:1043–47.

———. 1970. Predator-prey relationships and predation rates for crustacean zooplankters from some lakes in Western Canada. *Can. J. Zool.* 48:1229–40.

———. 1971. Crustacean plankton of 146 alpine and subalpine lakes and ponds in Western Canada. *J. Fish. Res. Bd.*, Canada, 28:311–21.

———. 1972. Zooplankton composition and change in an alpine lake. *Verh. Internat. Verein. Limnol.* 18:264–68.

———. 1975. Zooplankton and phytoplankton studies in Waterton Lakes, Alberta Park, Canada. *Verh. Internat. Verein. Limnol.* 19:571–79.

Anderson, R. S., and Raasveldt, L. G. 1974. *Gammarus* and *Chaoborus* predation. *Canadian Wildlife Serv. Occasional Paper* 18:1–23.

Archibald, C. P. 1975. Experimental observations on the effects of predation by goldfish (*Carassius auratus*) on the zooplankton of a small saline lake. *J. Fish. Res. Bd.*, Canada 32:1589–94.

Arcifa, M. S. 1976. A preliminary investigation on the cyclomorphosis of *Daphnia gessneri* Herbst, 1967, in a Brazilian reservoir. *Bolm. Zool. Univ. S. Paulo* 1:147–60.

Armstrong, R. A., and McGehee, R. 1976. Coexistence of species competing for shared resources. *Theor. Pop. Biol.* 9:317–28.

Arthur, D. K. 1976. Food and feeding of larvae of three fishes occurring in the California Current, *Sardinops sagax, Engraulis mordax* and *Trachurus symmetricus*. *Fish. Bull.*, U.S., 74:517–30.

Avery, R. A. 1968. Food and feeding relationship of three species of *Triturus* (Amphibia, Urodela) during the aquatic phases. *Oikos* 19:408–12.

Bainbridge, V. 1958. Some observations on *Evadne nordmanni* Loven. *J. Biol. Assn.*, U.K., 37:349–70.

Banse, K. 1964. On the vertical distribution of zooplankton in the sea. *Prog. Oceanog.* 2:53–125.

Bayly, I. A. E. 1963. Reversed diurnal vertical migration of planktonic crustacea in inland waters of low hydrogen ion concentration. *Nature* 200:704–C5.

Bayly, I. A. E., and Williams, W. A. 1973. *Inland waters and their ecology.* London: Longmans.

Beauchamp, P. de. 1952*a*. Un facteur de la variabilite chez les rotifères du genre *Brachionus*. *C.r. hebd. Séanc. Acad. Sci.*, Paris, 234:573–75.

———. 1952*b*. Variation chez les rotifères du genre *Brachionus C.r. heb. Séanc. Acad. Sci.*, Paris, 235:1355–56.

Beeton, A. M. 1960. The vertical migration of *Mysis relicta* in Lakes Huron and Michigan. *J. Fish. Res. Bd.*, Canada, 17:517–39.

Begg, G. W. 1976. The relationship between the diurnal movements of some of the zooplankton and the sardine *Limnothrissa miodon* in Lake Kariba, Rhodesia. *Limnol. and Oceanog.* 21:521–39.

Berg. A., and Grimaldi, E. 1966. Ecological relationships between planktophagic fish species in the Lago Maggiore. *Verh. Internat. Verein. Limnol.* 16:1065–73.

Beukema, J. J. 1968. Predation by the three-spined stickleback (*Gasterosteus aculeatus L.*): the influence of hunger and experience. *Behaviour* 31:1–126.

Birch, L. C. 1957. The meanings of competition. *Amer. Nat.* 91:5–18.

Blaxter, J. H. S. 1966. The effect of light intensity on the feeding ecology of herring. In R. Bainbridge, G. C. Evans, and O. Rackham, eds., Light as an ecological factor. *Brit. Ecol. Soc. Symp.* 6:393–409.

Blaxter, J. H. S., and Holliday, F. G. T. 1963. The behavior and physiology of herring and other clueipids. In F. S. Russel, ed., *Advances in marine biology*. New York: Academic Press, 1:261–393.

Bohmann, L., Engländer, H., Forese, H., Körner, I., Röckl, K. W., Schanzer, W., Schröder, W., and Kiang, H. 1940. Untersuchungen über die Ertragsgähigkeit einiger Seen Oberbayerns. *Int. Rev. ges. Hydrobiol.* 39:547–99.

Bosch, H. F., and Taylor, W. R. 1973*a*. Diurnal vertical migration of an esturaine cladoceran, *Podon polyphemoides*, in the Chesapeake Bay. *Mar. Biol.* 19:172–82.

———. 1973*b*. Distribution of the Cladoceran *Podon polyphemoides* in the Chesapeake Bay. *Mar. Biol.* 19:161–71.

Boulet, P. C. 1958. Contribution a l'étude experimentale de la perception visuelle du mouvement chez la perche et la seiche. *Mem. Mus. Natl. Hist. Nat.*, Paris, *Ser. A. Zool.* 17.

Bowers, J. A., and Grossnickle, N. E. 1978. The herbivorous habits of *Mysis relicta* in Lake Michigan. *Limnol. and Oceanog.* 23:767–76.

Boyd, C. M. 1976. Selection of particle sizes by filter-feeding copepods: A plea for reason. *Limnol. and Oceanog.* 21:175–80.

Bradbury, J. P. 1971. Paleolimnology of Lake Texcoco, Mexico. Evidence from diatoms. *Limnol. and Oceanog.* 16:180–200.

Brandl, Z., and Fernando, C. H. 1975. Investigations on the feeding of carnivorous cyclopoids. *Verh. Internat. Verein. Limnol.* 19:2959–65.

Braum, E. 1963. Die ersten Beutetang handlungen junger Blaufelchen (*Coregonus wartmanni* Bloch) und Hechte (*Esox lucius* L.). *Z. Tierpsychol.* 20:247–66.

Breen, P. A., and Mann, K. H. 1976. Destructive grazing of kelp by sea urchins in Eastern Canada. *J. Fish. Res. Bd.*, Canada, 33:1278–83.

Brett, J. R., and Groot, C. 1963. Some aspects of olfactory and visual responses in Pacific salmon. *J. Fish. Res. Bd.*, Canada, 20:287–303.

Brooks, J. L. 1946. Cyclomorphosis in *Daphnia*. *Ecol. Monogr.* 16:409–47.

———. 1957. The systematics of North American *Daphnia*. *Conn. Acad. Arts. and Sci. Mem.* 13.

———. 1959. Cladocera. In W. T. Edmondson, ed., *Freshwater biology*. New York: John Wiley, pp. 587–656.

———. 1965. Predation and relative helmet size in cyclomorphic *Daphnia*. *Proc. Nat. Acad. Sci.*, U.S., 53:119–26.
———. 1968. The effects of prey size selection by lake planktivores. *Syst. Zool.* 17:272–91.
———. 1969. Eutrophication and changes in the composition of zooplankton. In *Eutrophication, causes, consequences, correctives*. Washington, D.C.: National Academy of Sciences, 1970, pp. 236–55.
———. 1972. Extinction and the origin of organic diversity. *Trans. Conn. Acad. Arts and Sci.* 44:19–56.
Brooks, J. L., and Dodson, S. I. 1965. Predation, body size, and composition of plankton. *Science* 150:28–35.
Burbidge, R. G. 1974. Distribution, growth, selective feeding, and energy transformation of young-of-the-year blueback herring, *Alosa aestivalis* (Mitchill), in the James River, Virginia. *Trans. Amer. Fish. Soc.* 103:297–311.
Burckhardt, G. 1944. Verarmung des Planktons in kleinen Seen durch *Heterocope*. *Schweizz. Hydrol.* 9:121–24.
Burger, W. L., 1950. Novel aspects of the life history of two *Ambystomas*. *J. Tenn. Acad. Sci.* 25:252–57.
Burgis, M. J. 1967. A quantitative study of reproduction in some species of *Ceriodaphnia*. *J. of Animal Ecol.* 36:61–75.
———. 1973. Observations on the Cladocera of Lake George, Uganda. *J. of Zool. of London* 170:339–49.
Burns, C. W. 1968*a*. The relationship between body size of filter-feeding cladocera and the maximum size of particle ingested. *Limnol. and Oceanog.* 13(4):675–78.
———. 1968*b*. Progress report to the Trustees of the Pond Mountain Natural Area, unpublished.
———. 1969*a*. Particle size and sedimentation in the feeding behavior of two species of *Daphnia*. *Limnol. and Oceanog.* 14(3):392–402.
———. 1969*b*. Relation between filtering rate, temperature, and body size in four species of *Daphnia*. *Limnol. and Oceanog.* 14(5):693–700.
Butorina, L. G. 1971. O sposo benosti *Polyphemus pediculus* (L) pitatsja bakterijami i prostejšimi. *Biol. vnutr. vod. Inform. bjul.* 11:47–48.
Calef, G. W. 1973. Natural mortality of tadpoles in a population of *Rana aurora*. *Ecology* 54:741–58.
Chace, F. A., Mackin, J. G., Hubricht, L., Banner, A. H., Hobbs, H. H. 1959. Malacostraca. In W. T. Edmondson, ed., *Freshwater biology*. New York: J. Wiley, pp. 869–901.

Charnov, E. L. 1973. Optimal foraging: some theoretical explorations. Ph.D. thesis, University of Washington, Seattle.

Chesson, J. 1978. Measuring preference in selective predation. *Ecology* 59:211–15.

Clark, A. S., and Carter, J. C. H. 1974. Population dynamics of cladocerans in Sunfish Lake, Ontario. *Can. J. Zool.* 52:1235–42.

Cody, M. L. 1974. *Competition and the structure of bird communities.* Princeton: Princeton University Press.

Confer, J. L. 1971. Intra-zooplankton predation by *Mesocyclops edax* at natural prey densities. *Limnol. and Oceanog.* 16:663–66.

Confer, J. L., and Blades, P. I. 1975. Omnivorous zooplankton and planktivorous fish. *Limnol. and Oceanog.* 20:571–79.

Confer, J. L., Howick, G. L., Corzette, M. H., Kramer, S. L., Fitzgibbon, S., and Landesberg, R. 1978. Visual predation by planktivores. *Oikos* 31:27–37.

Connell, J. H. 1975. Some mechanisms producing structure in natural communities. In M. L. Cody and J. M. Diamond, eds., *Ecology and evolution of communities.* Cambridge: Harvard University Press, Belknap Press, pp. 460–90.

Conover, R. J. 1968. Zooplankton-life in a nutritionally dilute environment. *Amer. Zool.* 8:107–18.

Costa, R. R., and Cummins, K. W. 1972. The contribution of *Leptodora* and other zooplankton to the diet of various fish. *Amer. Mid. Nat.* 87:559–64.

Covich. A. 1972. Ecological economics of seed consumption by *Peromyscus. Trans. Conn. Acad. Arts and Sci.* 44:71–93.

Cowgill, U. M., Goulden, C. E., Hutchinson, G. E., Patrick, R., Răcek, A. A., and Tsukada, M. 1966. The history of Laguna de Petenxil. *Conn. Acad. Arts and Sci.* 17.

Cremer, G. A., and Duncan, A. 1969. A seasonal study of zooplanktonic respiration under field conditions. *Verh. Internat. Verein. Limnol.* 17:181–90.

Croze, H. 1970. Searching image in carrion crows. *Z. Tierpsychol.* Suppl. 5.

Cummins, K. W., Costa, R. R., Rowe, R. E., Moshiri, G. A., Scanlan, R. M., and Zajdel, R. K. 1969. Ecological energetics of a natural population of the predaceous zooplankter *Leptodora kindtii* Focke (Cladocera). *Oikos* 20:189–223.

Curio, E. 1976. *The ethology of predation.* Berlin: Springer-Verlag.

Daborn, G. R. 1975. Life history and energy relations of the giant fairy shrimp *Branchinecta gigas Lynch 1937 (Crustacea: Anostraca). Ecology* 56:1025–39.

D'Angelo, S. A., Gordon, A. S., and Charipper, H. A. 1941. The role of the thyroid and putuitary glands in the anomalous effect of inanition on amphibian metamorphosis. *J. Exp. Zool.* 87:259–N77.

Darwin, C. 1859. *On the origin of species*, facsimile of 1st ed., ed. E. Mayr. London and Cambridge, Mass.: Murray and Harvard University Press. 1964.

David, P. M. 1961. The influence of vertical migration on speciation in the oceanic plankton. *Syst. Zool.* 10:10–16.

Davis, M. B. 1969. Climatic changes in southern Connecticut recorded by pollen deposition at Rogers Lake. *Ecology* 50:409–22.

Dawkins, M. 1971. Perceptual changes in chicks: another look at the "search image" concept. *Anim. Behav.* 19:556–74.

De Bernardi, R., and Guissani, G. 1975. Population dynamics of three cladocerans of Lago Maggiore related to predation pressure by a planktophagous fish. *Verh. Internat. Verein. Limnol.* 19:2906–12.

Deevey, E. S., and Deevey, G. B. 1971. The American Species of *Eubosmina* Seligo (Crustacea, Cladocera). *Limnol. and Oceanog.* 16:201–18.

Dent, J. N. 1968. Survey of amphibian metamorphosis. In W. Etkin and L. I. Gilbert, *Metamorphosis; a problem in developmental biology*. New York: Appleton-Century-Crofts, pp. 271–311.

Denton, E. J. 1971. Reflectors in fishes. *Sci. Am.* 224:64–72.

Denton, E. J., and Nichol, J. A. C. 1963. Why fish have silvery sides; and a method of measuring reflectivity. *J. Physiol.*, London, 165:13–15.

Dexter, R. W. 1959. Anostraca. In W. T. Edmondson, ed., *Freshwater biology*. New York: Wiley, pp. 558–71.

Dickman, M. 1968. The effect of grazing by tadpoles in the structure of a periphyton community. *Ecology* 49:1188–90.

Dodson, S. I. 1970. Complementary feeding niches sustained by size-selective predation. *Limnol. and Oceanog.* 15:131–37.

———. 1972. Mortality in a population of *Daphnia rosea*. *Ecology* 53:1011–23.

———. 1974*a*. Zooplankton competition and predation: an experimental test of the size-efficiency hypothesis. *Ecology* 55:605–13.

———. 1974*b*. Adaptive change in plankton morphology in response to size-selective predation: A new hypothesis of cyclomorphosis. *Limnol. and Oceanog.* 19:721–29.

Dodson, S. I., and Dodson, V. E. 1971. The diet of *Ambystoma tigrinum* larvae from Western Colorado. *Copeia* 1971:614–24.

Dodson, S. I., Edwards, C., Wiman, F., and Normandin, J. C. 1976. Zooplankton: specific distribution and food abundance. *Limnol. and Oceanog.* 21:309–13.

Drenner, R. W., Strickler, J. R., and O'Brien, W. J. 1978. Capture probability: The role of zooplankter escape in the selective feeding of planktivorous fish. *J. Fish. Res. Bd.* Canada, 34:1370–73.

Dühr, B. 1955. Uber Bewegung, Orientierung und Beutefang der Corethralarve (*Chaoborus crystallinus* De Geer). *Zool. Jahrb. Physiol.* 65:387–429.

Dumont, H. J. 1972. A competition-based approach to the reverse vertical migration in zooplankton and its implications, chiefly based on a study of the interactions of the rotifer *Asplanchna priodonta* Gosse with several crustacea entomostraca. *Int. Rev. ges. Hydrobiol.* 57:1–38.

Duncan, A. 1975. Production and biomass of three species of *Daphnia* coexisting in London reservoirs. *Verh. Internat. Verein. Limnol.* 19:2858–67.

Dzyuban, N. A. 1939. New data on the feeding of some cyclopidae. Trudy Mosk Tekh Inst Rybnogo Khoz i Prom imeni. A. I. Mikoyana, pp. 163–72. Translated by Canada FBB, Nanaimo Station CTS, March 1964, p. 79.

Edmondson, W. T. 1959. Rotifera. In W. T. Edmondson, ed., *Freshwater biology*. New York: Wiley, pp. 420–94.

Edmondson, W. T., and Winberg, G. G., eds. 1971. *A manual on methods for the assessment of secondary productivity in fresh waters.* IBP Handbook No. 17. Oxford: Blackwell.

Edmunds, M. 1974. *Defence in animals*. New York: Longman.

Edwards, J. S. 1963. Arthropods as predators. In J. D. Carthy and C. L. Duddington, eds., *Viewpoints in biology*. Washington, D.C.: Butterworth, 2:85–114.

Eggers, D. M. 1976. Theoretical effect of schooling by planktivorous fish predators on rate of prey consumption. *J. Fish. Res. Bd.*, Canada, 33:1964–71.

———. 1977. The nature of prey selection by planktivorous fish. *Ecology* 58:46–59.

———. 1978. Limnetic feeding behavior of juvenile sockeye salmon in Lake Washington and predator avoidance. *Limnol. and Oceanog.* 23:1114–25.

Egloff, D. A., and Palmer, D. S. 1971. Size relations of the filtering area of two *Daphnia* species. *Limnol. and Oceanog.* 16:900–05.

Elton, C. 1927. *Animal ecology.* London: Sidgwick & Jackson.

Estes, J. A., and Palmisano, J. F. 1974. Sea otters: Their role in structuring nearshore communities. *Science* 185:1058–60.

Etkin, W., and Gilbert, L. I., eds. 1968. *Metamorphosis; a problem in developmental biology*. New York: Appleton-Century-Crofts.

Fedorenko, A. Y. 1975*a*. Instar and species specific diets in two species of *Chaoborus*. *Limnol. and Oceanog.* 20 (2):238–49.

———. 1975*b*. Feeding characteristics and predation impact of *Chaoborus* (Diptera, Chaoboridae) larvae in a small lake. *Limnol. and Oceanog.* 20 (2):250–58.

Fedorenko, A. Y., and Swift, M. C. 1972. Comparative biology of *Chaoborus americanus* and *Chaoborus trivittatus* in Eunice Lake, British Columbia. *Limnol. and Oceanog.* 17:721–30.

Feigenbaum, D., and Reeve, M. R. 1977. Prey detection in the Chaetognatha: Response to a vibrating probe and experimental determination of attack distance in large aquaria. *Limnol. and Oceanog.* 22:1052–58.

Fleminger, A., and Clutter, R. I. 1965. Avoidance of towed nets by zooplankton. *Limnol. and Oceanog.* 10:96–104.

Fox, H. M. 1948. The haemoglobin of *Daphnia*. *Proc. R. Soc., Ser. 135:195–212.*

Frank, P. W. 1957. Coactions in laboratory populations of two species of *Daphnia*. *Ecology* 38:510–19.

Frost, B. W. 1972. Effects of size and concentration of food particles on the feeding behavior of the marine planktonic copepod *Calanus pacificus*. *Limnol. and Oceanog.* 17:805–15.

———. 1974. A threshold feeding behavior in *Calanus pacificus*. *Limnol. and Oceanog.* 20:263–66.

———. 1977. Feeding behavior of *Calanus pacificus* in mixtures of food particles. *Limnol. and Oceanog.* 22:472–91.

Fryer, G. 1957. The feed mechanism of some freshwater cyclopoid copepods. *Proc. Zool. Soc.*, London, 129:1–25.

Fryer, G., and Iles, T. D. 1972. The cichlid fishes of the great lakes of Africa: their biology and evolution. Edinburgh: Oliver and Boyd.

Fuchs, E. H. 1967. Life history of the emerald shiner, *Notropis atherinoides*, in Lewis and Clark Lake, South Dakota. *Trans. Amer. Fish. Soc.* 96:247–56.

Galbraith, M. G., Jr. 1967. Size-selective predation on *Daphnia* by rainbow trout and yellow perch. *Trans. Amer. Fish. Soc.* 96:1–10.

Gerritsen, J., and Strickler, J. R. 1977. Encounter probabilities and community structure in zooplankton: a mathematical model. *J. Fish. Res. Bd.*, Canada, 34:73–82.

Gilbert, J. J. 1967. *Asplanchna* and postero lateral spine production in *Brachionus calyciflorus. Arch. Hydrobiol.* 64:1–62.

———. 1973*a*. Induction and ecological significance of gigantism in the rotifer *Asplanchna sieboldi. Science* 181:63–66.

———. 1973*b*. The adaptive significance of polymorphism in the rotifer *Asplanchna*. Humps in males and females. *Oecologia* 13:135–46.

Gilbert, J. J., and Thompson, G. A. 1968. Alpha tocopherol control of sexuality and polymorphism in the rotifer *Asplanchna. Science* 159:734–6.

Gilbert, J. J., and Waage, J. K. 1967. *Asplanchna, asplanchna* substance, and postero lateral spine length variation of the rotifer *Brachionus calyciflorus* in a natural environment. *Ecology* 48:1027–31.

Gliwicz, Z. M. 1969. The share of algae, bacteria and trypton in the food of the pelagic zooplankton of lakes with various trophic characteristics. *Bull. L'Acad. Polonaise Sci.* 17:159–65.

———. 1977. Food size selection and seasonal succession of filter feeding zooplankton in an eutrophic lake. *Ekol. Pol.* 25:179–225.

Gliwicz, Z. M., and Biesiadka, E. 1975. Pelagic water mites (Hydracarina) and their effect on the plankton community in a neotropical manmade lake. *Arch. Hydrobiol.* 76:65–88.

Glynn, P. W. 1979. Coral reef growth in the Galápagos: Limitation by Sea Urchins. *Science* 203:47–49.

Goldman, C. R., Morgan, M. D., Threlkeld, S. T., and Angell, N. 1979. A population dynamics analysis of the cladoceran disappearance from Lake Tahoe, California-Nevada. *Limnol. and Oceanog.* 24:289–97.

Goldspink, C. R., and Scott, D. B. C. 1971. Vertical migration of *Chaoborus flavicans* in a Scottish loch. *Freshwater Biol.* 1:411–21.

Gophen, M. 1979. Extinction of *Daphnia lumholtzi* (Sars) in Lake Kinneret (Israel). *Aquaculture* 16:67–71.

Gould, S. J. 1977. *Ontogeny and phylogeny*. Cambridge: Harvard University Press, Belknap Press. 501 p.

Goulden, C. E. 1971. Environmental control of the abundance and distribution of the chydorid Cladocera. *Limnol. and Oceanog.* 16:320–31.

Green, J. 1967. The distribution and variation of *Daphnia lumholtzii* (Crustacea: Cladocera) in relation to fish predation in Lake Albert, East Africa. *J. Zool.*, London, 151:181–97.

———. 1971. Associations of Cladocera in the zooplankton of the lake sources of the White Nile. *J. Zool.*, London, 165:373–414.

———. 1974. *Asplanchna* and the spines of *Brachionus calyciflorus* in two Javanese sewage ponds. *Freshwater Biol.* 4:223–26.

Greenwood, P. H. 1953. Feeding mechanisms of the cichlid fish, *Tilapia esculenta* Graham. *Nature* 172:207–08.

Greze, V. N. 1963. The determination of transparency among planktonic organisms and its protective significance. *Dokl. Akad. Nauk. SSSR.* (English trans.) 151(2):956–58.

Hairston, N. G., Jr. 1976. Photoprotection by carotenoid pigments in the copepod *Diaptomus nevadensis. Proc. Nat. Acad. Sci.,* U.S., 73:971–74.

———. 1977. The adaptive significance of carotenoid pigmentation in *Diaptomus* (copepoda). Ph.D. thesis, University of Washington, Seattle.

Halbach, U. 1971. Zum adaptivwert de zyklomorphen Dornenbildung von *Brachionus calyciflorus* Pallas (Rotatoria). *Oecologia* 6:267–88.

Hall, D. J. 1964. An experimental approach to the dynamics of a natural population of *Daphnia galeata mendotae. Ecology* 45:94–111.

Hall, D. J., Cooper, W. E., and Werner, E. E. 1970. An experimental approach to the production dynamics and structure of freshwater animal communities. *Limnol. and Oceanog.* 15:839–928.

Hall, D. J., Threlkeld, S. T., Burns, C. W., and Crowley, P. H. 1976. The size-efficiency hypothesis and the size structure of zooplankton communities. *Ann. Rev. Ecol. Syst.* 7:177–208.

Haney, J. F. 1973. An in situ examination of the grazing activities of natural zooplankton communities. *Arch. Hydrobiol.* 72:87–132.

Haney, J. F., and Hall, D. J. 1975. Diel vertical migration and filter-feeding activities of *Daphnia. Arch. Hydrobiol.* 75:413–41.

Hardy, A. C. 1956. *The open sea; its natural history*. New York: Collins.

Harris, J. E., and Mason, P. 1956. Vertical migration in eyeless *Daphnia. Proc. R. Soc., Ser. B* 145:280–90.

Hebert, P. D. N. 1974. Enzyme variability in natural populations of *Daphnia magna* I. Population structure in East Anglia. *Evolution* 28:546–56.

———. 1977. A revision of the taxonomy of the genus *Daphnia* (Crustacea: Daphnidae) in Southeastern Australia. *Aust. J. Zool.* 25:371–98.

Heisey, O., and Porter, K. G. 1977. The effect of ambient oxygen concentration on filtering and respiration rates of *Daphnia galeata mendotae* and *Daphnia magna. Limnol. and Oceanog.* 22:839–45.

Hemmings, C. C. 1966. Factors influencing the visibility of objects underwater. In R. Bainbridge, G. C. Evans, and O. Rackham, eds., Light as an ecological factor. *Brit. Ecol. Soc. Symp.* 6:359–74.

Hester, F. J. 1968. Visual contrast thresholds of the goldfish (*Carassius auratus*). *Vision Res.* 8:1315–36.

Heyer, W. R., McDiarmid, R. W., and Weigmann, D. L. 1975. Tadpoles, predation and pond habitats in the tropics. *Biotropica* 7(2):100–11.

Hillbricht-Ilkowska, A., and Karabin, A. 1970. An attempt to estimate consumption, respiration, and production of *Leptodora kindtii* (Focke) in field and laboratory experiments. *Pol. Arch. Hydrobiol.* 17:81–86.

Holling, C. S. 1959. The components of predation, as revealed by a study of small mammal predation of the European pine sawfly. *Can. Entomol.* 91:293–332.

———. 1966. The functional response of invertebrate predators to prey density. *Mem. Entomol. Soc. Can.* 48:1–86.

Holst, E. V. 1948. Quantitative Untersuchungen über Umstimmungsvorgängeim Zentralnerven system I. Der Einfluss des Appetits auf das Gleichgewichtsverhalten bei *Pterophyllum*. *Z. Vergl. Physiol.* 31:134–48.

Hrbáček, J. 1959. Circulation of water as a main factor influencing the development of helmets in *Daphnia cucullata* Sars. *Hydrobiologia* 13:170–85.

———. 1962. Species composition and the amount of zooplankton in relation to the fish stock. *Rozpr. ČSAV, Ser. mat. nat. sci.* 72:1–117.

———. 1977. Competition and predation in relation to species composition of freshwater zooplankton, mainly *Cladocera*. In J. Cairns, Jr., ed., *Aquatic microbial communities*, 305–53. Garland Publishing, New York and London.

Hrbáček, J., Dvořakova, M., Kořínek, V., and Procházkóva, L. 1961. Demonstration of the effect of the fish stock on the species composition of zooplankton and the intensity of metabolism of the whole plankton association. *Verh. Internat. Verein. Limnol.* 14:192–95.

Huffaker, C. B. 1958. Experimental studies on predation: dispersion fac-

tors and predator-prey oscillations. *Hilgardia* 27:343–83.
Hunter, J. R. 1979. The feeding behavior and ecology of marine fish larvae. In J. E. Bardach, ed., *The physiological and behavioral manipulation of food fish as production and management tools.*
Hunter, J. R., and Thomas, G. L. 1974. Effect of prey distribution and density on the searching and feeding behavior of larval anchovy *Engraulis mordax*, Girard. In J. H. S. Blaxter, ed., *The early life history of fish*. Berlin: Springer-Verlag, pp. 559–74.
Hurlbert, S. H., Mulla, M. S. and Willson, H. R. 1972*a*. Effects of an organophosphorous insecticide on the phytoplankton, zooplankton, and insect populations of freshwater ponds. *Ecol. Monogr.* 42:269–99.
Hurlbert, S. H., Zedler, J. and Fairbanks, D. 1972*b*. Ecosystem alteration by mosquitofish (*Gambusia affinis*) predation. *Science* 175:639–41.
Hutchinson, B. P. 1971. The effect of fish predation on the zooplankton of ten Adirondack lakes, with particular reference to the alewife, *Alosa pseudoharengus*. *Trans. Amer. Fish. Soc.* 100:325–35.
Hutchinson, G. E. 1951. Copepodology for the ornithologist. *Ecology* 32:571–77.
———. 1957. Concluding remarks. *Cold Spring Harbor Symp. Quant. Biol.* 22:415–27.
———. 1959. *A treatise on limnology*, 3 vols. Vol. 1: *Geography, physics and chemistry*. New York: Wiley.
———. 1967. *A treatise on limnology*, 3 vols. Vol. 2: *Introduction to lake biology and the limnoplankton*. New York: Wiley.
———. 1975. *A treatise on limnology*, 3 vols. Vol. 3: *Limnological botany*. New York: Wiley.
Ingle, D. 1968. Spatial dimensions of vision in fish. In D. Ingle, ed., *The central nervous system and fish behavior*. Chicago: University of Chicago Press, pp. 51–59.
Ivlev, V. S. 1961. *Experimental ecology of the feeding of fishes*. New Haven, Conn.: Yale University Press.
Jacobs, J. 1965. Significance of morphology and physiology of *Daphnia* for its survival in predatory-prey experiments. *Naturwissenschaften* 52:141.
———. 1967. Untersuchen zuer Funktion and Evolution der Zyklomorphose bei *Daphnia*, mit besonderer Berücksichtigung der Selektion durch Fische. *Arch. Hydrobiol.* 62:467–541.
———. 1974. Qualitative measurements of food selection. *Oecologia* 14:413–17.

James, M. T. 1959. Diptera. In W. T. Edmondson, ed., *Freshwater biology*. New York: Wiley, pp. 1057–59.

Janssen, J. 1976. Feeding modes and prey size selection in the alewife (*Alosa pseudoharengus*). *J. Fish. Res. Bd*, Canada, 33:1972–75.

———. 1978. Feeding-behavior repertoire of the alewife, *Alosa pseudoharengus*, and the ciscoes *Coregonus hoyi* and *C. artedii*. *J. Fish. Res. Bd.*, Canada, 35:249–53.

Johnson, D. M. 1973. Predation by damselfly naiads on cladoceran populations: fluctuating intensity. *Ecology* 54:251–68.

Johnson, D. M., Akre, B. G., and Crowley, P. H. 1975. Modeling arthropod predation: wasteful filling by damselfly naiads. *Ecology* 56:1081–93.

Johnstone, K. 1972. The vanishing harvest. *The Canadian fishing crisis*. Montreal: *Montreal Star*.

June, F. C., and Carlson, F. T. 1971. Food of young atlantic menhaden, *Brevoortia tyrannus*, in relation to metamorphosis. *Fish. Bull.*, U.S., 68:493–512.

Katona, S. K. 1973. Evidence for sex pheromones in planktonic copepods. *Limnol. and Oceanog*. 18:574–83.

Keast, A. 1970. Food specializations and bioenergetic interrelations in the fish faunas of some small Ontario waterways. In J. H. Steel, ed., *Marine food chains*. Edinburgh: Oliver and Boyd, pp. 377–411.

Kerfoot, W. C. 1974*a*. Egg-size cycle of a cladoceran. *Ecology* 55:1259–70.

———. 1974*b*. New accumulation rates and the history of cladoceran communities. *Ecology* 55:51–61.

———. 1975*a*. Seasonal changes of *Bosmina* (Crustacea: Cladocera) in Frains Lake, Michigan: Laboratory observations of phenotypic changes induced by inorganic factors. *Freshwater Biol*. 5:227–43.

———. 1975*b*. The divergence of adjacent populations. *Ecology* 56:1298–1313.

———. 1977*a*. Implications of copepod predation. *Limnol. and Oceanog*. 22:316–25.

———. 1977*b*. Competition in cladoceran communities: the cost of evolving defenses against copepod predation. *Ecology* 58:303–13.

Kime, J. B. 1974. Ecological relationships among three species of aeshnid dragonfly larvae (Odonate: Aeshnidae). Ph.D. thesis, University of Washington, Seattle.

Kislalioglu, M. 1976. Prey "handling time" and its importance in food selection by the 15-spined stickleback, *Spinachia spinachia* L. *J. Exp.*

Mar. Biol. Ecol. 25:151–58.

Kislalioglu, M., and Gibson, R. N. 1976. Some factors governing prey selection by the 15-spinel stickleback, *Spinachia spinachia* L. *J. Exp. Mar. Biol. Ecol.* 25:159–69.

Kliewer, E. V. 1970. Gillraker variation and diet in lake whitefish *Coregonus clupeaformis* in northern Manitoba. In C. C. Lindsey and C. S. Woods, eds., *Biology of coregonid fishes.* Winnipeg: University of Manitoba Press, pp. 147–65.

Korínek, V. 1972. Results of the study of some links of the food chain in a carp pond in Czechoslovakia. In Z. Kajak and Z. Hillbricht-Ilkowska, eds., *Productivity problems of freshwaters.* Warsaw-Cracow: pp. 541–53.

Kozhov, M. 1963. *Lake Baikal and its life.* The Hague: W. Junk Publishers.

Krebs, J. R. 1973. Behavioral aspects of predation. In P. P. G. Bateson and P. H. Klopfer, eds., *Perspectives in ethology.* New York: Plenum Press, 1:73–111.

Kring, R. L., and O'Brien, W. J. 1976. Effect of varying oxygen concentrations on the filtering rate of *Daphnia* pulex. *Ecology* 57:808–14.

Kutkuhn, J. H. 1957. Utilization of plankton by juvenile gizzard shad in a shallow prairie lake. *Trans. Amer. Fish. Soc.* 87:80–103.

Kwik, J. K., and Carter, J. C. H. 1975. Population dynamics of limnetic cladocera in a beaver pond. *J. Fish. Res. Bd.*, Canada, 32:341–46.

Landry, M. R. 1978. Predatory feeding behavior of a marine copepod, *Labidocera trispinosa. Limnol. and Oceanog.* 23:1103–13.

Lane, P. A. 1975. The dynamics of aquatic systems: a comparative study of the structure of four zooplankton communities. *Ecol. Monogr.* 45:307–36.

LaRow, E. J. 1970. The effect of oxygen tension on the vertical migration of *Chaoborus* larvae. *Limnol. and Oceanog.* 15:357–62.

Lasenby, D. C., and Langford, R. R. 1973. Feeding and assimilation of *Mysis relicta. Limnol. and Oceanogr.* 18:280–85.

Lasker, R., and Zweifel, J. R. 1978. Growth and survival of first-feeding northern anchovy larval (*Engraulis mordax*) in patches containing different proportions of large and small prey. In J. H. Steel, ed., *Spatial pattern in plankton communities* IV:3. New York and London: Plenum Press, pp. 329–54.

Leong, R. J. H., and O'Connell, C. P. 1969. A laboratory study of particulate and filter feeding of the northern anchovy (*Engraulis mordax*). *J.*

Fish. Res. Bd., Canada, 26:557–82.

Levin, S. A., and Segel, L. A. 1976. Hypothesis for origin of planktonic patchiness. *Nature* 259:659.

Lewis, W. J., Jr. 1975. Distribution and feeding habits of a tropical *Chaoborus* population. *Verh. Internat. Verein. Limnol.* 19:3106–19.

Lindstrom, T. 1955. On the relation fish-size–food-size. *Rept. Inst. Freshwater Res.*, Drottningholm, 36:133–47.

Losos, B., and Hetesa, J. 1973. The effect of mineral fertilization and of carp fry and the composition and dynamics of plankton. *Hydrobiol. Stud.* 3:173–217.

Lynch, M. 1977. Fitness and optimal body size in zooplankton populations. *Ecology* 58:763–74.

———. 1979. Interacting resources and accommodating species. *Amer. Nat.* 114.

Lythgoe, J. N. 1966. Visual pigments and underwater vision. In R. Bainbridge, G. C. Evans, and O. Rackham, eds., Light as an ecological factor. *Brit. Ecol. Soc. Symp.* 6:375–91.

Macagno, E. R. 1978. Mechanism for the formation of synaptic projections in the arthropod visual system. *Nature* 275:318–20.

Macan, T. T. 1977. The influence of predation on the composition of freshwater animal communities. *Biol. Rev.* 52:45–70.

MacArthur, R. H. 1972. *Geographical ecology, patterns in the distribution of species*. New York: Harper & Row.

Main, R. A. 1962. The life history and food relations of *Epischura lacustris* Forbes (Copepoda: Calanoida). Ph.D. thesis, University of Michigan.

Maiorana, V. C. 1977. Density and competition among sunfish: some alternatives. *Science* 195:94–95.

Makarewicz, J. C., and Likens, G. E. 1975. Niche analysis of a zooplankton community. *Science* 190:1000–03.

Maly, E. J. 1978. Some factors influencing size of *Diaptomus shoshone*. *Limnol. and Oceanog.* 23:835–37.

Marshall, S. M., and Orr, A. P. 1964. Grazing by copepods in the sea. In *Grazing in Terrestrial and Marine Environments*. Oxford: Blackwell, pp. 227–38.

———. 1972. *The biology of a marine copepod*. London: Oliver and Boyd.

May, R. C. 1970. Feeding larval marine fishes in the laboratory: a review. *Calif. Coop. Oceanic Fish. Invest. Rept.* 14:76–83.

McLaren, I. A. 1963. Effects of temperature on growth of zooplankton and the adaptive value of vertical migration. *J. Fish. Res. Bd.*, Canada, 20:685–727.

———. 1974. Demographic strategy of vertical migration by a marine copepod. *Amer. Nat.* 108:91–102.

McNaught, D. C. 1975. A hypothesis to explain the succession from calanoids to cladocerans during eutrophication. *Verh. Internat. Verein. Limnol.* 19:724–31.

McNaught, D. C., and Hasler, A. D. 1961. Surface schooling and feeding behavior in the white bass, *Roccus chrysops* (Rafinesque), in Lake Mendota. *Limnol. and Oceanog.* 6:53–60.

McPhail, J. D. 1969. Predation and the evolution of a stickleback (Gasterosteus). *J. Fish. Res. Bd.*, Canada, 26:3183–3208.

Mellors, W. K. 1975. Selective predation of ephippial *Daphnia* and the resistance of ephippial eggs to digestion. *Ecology* 56:974–80.

Menge, B. A., and Sutherland, J. P. 1976. Species diversity gradients: synthesis of the roles of predation, competition and temporal heterogeneity. *Amer. Nat.* 110:351–69.

Merret, N. R., and Roe, H. S. J. 1974. Patterns and selectivity in the feeding of certain mesopelagic fishes. *Mar. Biol.* 28:115–26.

Milinski, M., and Curio, E. 1975. Untersuchungen zur selektion durch Rauber gegen Vereinzelung der Heute. *Z. Tierpsychol.* 37:400–02.

Miracle, M. R. 1974. Niche structure in freshwater zooplankton: a principal components approach. *Ecology* 55:1306–16.

Monakov, A. V. 1972. Review of studies on feeding of aquatic invertebrates conducted at the Institute of Biology of Inland Waters, Academy of Science, U.S.S.R. *J. Fish. Res. Bd.*, Canada, 29:363–83.

Mordukhai-Boltovskaia, E. D. 1958. Preliminary notes on the feeding of the carnivorous Cladocerans *Leptodora kindtii* and *Bythotrephes. Dokl. Akad. SSSR Biol. Sci. Sect.* 122:828–30.

Moriarty, C. M., and Moriarty, D. J. W. 1973. Quantitative estimation of the daily ingestion of phytoplankton by *Tilapia nilotica* and *Haplochromis nigripinnis* in Lake George, Uganda. *J. Zool.*, London, 171:15–23.

Morowitz, H. 1968. *Energy flow in biology*. New York: Academic Press.

Moshiri, G. A., Cummins, K. W., and Costa, R. R. 1969. Respiratory energy expenditure by the predaceous zooplankter *Leptodora kindtii* (Focke) (Crustacea: Cladocera). *Limnol. and Oceanog.* 14:475–84.

Munz, F. W., and McFarland, W. N. 1973. The significance of spectral position in the rhodopsins of tropical marine fishes. *Vision Res.* 13(10):1829–74.

Murdoch, W. W. 1969. Switching in general predators: Experiments in predator specificity and stability of prey populations. *Ecol. Monogr.* 39(4):335–54.

———. 1971. The developmental response of predators to changes in prey density. *Ecology* 52:132–37.

Murdoch, W. W., Avery, S., and Smyth, M. E. B. 1975. Switching in predatory fish. *Ecology* 56:1094–1105.

Narver, D. W. 1970. Diel vertical movements and feeding of underyearling sockeye salmon and the limnetic zooplankton in Babine Lake, British Columbia. *J. Fish. Res. Bd.*, Canada, 27:281–316.

Neill, S. R., and Cullen, J. M. 1974. Experiments on whether schooling by their prey affects the hunting behavior of cephalopod and fish predators. *J. Zool.*, London, 172:549–69.

Neill, W. E. 1975. Experimental studies of microcrustacean competition, community composition and efficiency of resource utilization. *Ecology* 56:809–26.

Nikolsky, G. V. 1963. *The ecology of fishes.* New York and London: Academic Press.

Nilsson, N. A. 1960. Seasonal fluctuations in the food segregation of trout, char and whitefish in 14 North-Swedish lakes. Rept. *Inst. Freshwater Res.*, Drottningholm, 41:185–205.

———. 1963. Interaction between trout and char in Scandinavia. *Trans. Amer. Fish. Soc.* 92:276–85.

———. 1967. Interactive segregation between fish species. In S. D. Gerking, ed., *The biological basis of freshwater fish production.* Oxford: Blackwell, pp. 295–313.

———. 1978. The role of size-biased predation in competition and interactive segregation in fish. In S. D. Gerking, Ed., *Ecology of Freshwater Fish Production.* New York and Toronto: Halsted Press, Wiley, pp. 303–25.

Nilsson, N. A., and Pejler, B. 1973. On the relation between fish fauna and zooplankton composition in north Swedish lakes. *Inst. Freshwater Res.*, Drottningholm, 53:51–77.

Noble, R. L. 1975. Growth of young yellow perch (*Perca flavescens*) in relation to zooplankton populations. *Trans. Amer. Fish. Soc.* 104:731–41.

Oaten, A. 1977. Transit time and density-dependent predation on a patchly distributed prey. *Amer. Nat.* 111:1061–75.

O'Brien, W. J. 1975. Some aspects of the limnology of the ponds and lakes of the Noatak drainage basin, Alaska. *Verh. Internat. Verein. Limnol.* 19:472–79.

O'Brien, W. J., and Schmidt, D. 1979. Arctic *Bosmina* morphology and copepod predation. *Limnol. and Oceanog.* 23:1231–37.

O'Brien, W. J., Slade, N. A., and Vinyard, G. L. 1976. Apparent size as the determinant of prey selection by bluegill sunfish (*Lepomis macrochirus*). *Ecology* 57:1304–10.

O'Brien, W. J., and Vinyard, G. L. 1974. Comment on the use of Ivlev's electivity index with planktivorous fish. *J. Fish. Res. Bd.*, Canada, 31:1427–29.

O'Brien, W. J., and Vinyard, G. L. 1978. Polymorphism and predation: The effect of invertebrate predation on the distribution of two varieties of *Daphnia carinata* in South India ponds. *Limnol. and Oceanog.* 23:452–60.

O'Connell, C. P. 1972. The interrelation of biting and filtering in the feeding activity of the northern anchovy (*Engraulis* mordax). *J. Fish. Res. Bd.*, Canada, 29:285–93.

Ogden, J. C., Brown, R. A. and Salesky, N. 1973. Grazing by echinoid *Diadema antillarum* Philippi: Formation of halos around West Indian Patch Reefs. *Science* 182:715–16.

Orians, G. H., and Horn, H. S. 1969. Overlap in foods and foraging of four species of blackbirds in the potholes of central Washington. *Ecology* 50:930–38.

Orians, G. H., Charnov, E. L., and Hyatt, K. 1976. Ecological implications of resource depression. *Amer. Nat.* 110:247–59.

Paine, R. T. 1966. Food web complexity and species diversity. *Amer. Nat.* 100:65–75.

Paine, R. T., and Vadas, R. L. 1969. The effects of grazing by sea urchins, *Strongylocentrotus* spp. on benthic algal populations. *Limnol. and Oceanog.* 14:710–19.

Palmer, A. R. 1977. Function of shell sculpture in marine gastropods: hydrodynamic destabilization in *Ceratostoma foliatum*. *Science* 197:1293–95.

———. 1979. Fish predation and the evolution of gastropod shell sculpture: experimental and geographic evidence. *Evolution*, 33:697–713.

Parma, S. 1971. *Chaoborus flavicans* (Meigen) (Diptera, Chaoboridae): An autecological study. Ph.D. thesis, Groningen.

Pastorok, R. A. 1978. Predation by *Chaoborus* larvae and its impact on the zooplankton community. Ph.D. thesis, University of Washington, Seattle.

Patalas, K. 1972. Composition and horizontal distribution of Crustacean plankton in Lake Ontario. *J. Fish. Res. Bd.*, Canada, 26:2135–64.

———. 1975. The crustacean plankton communities of fourteen North American great lakes. *Verh. Internat. Verein. Limnol.* 19:504–11.

Paulson, D. R. 1973. Predator polymorphism and apostatic selection. *Evolution* 27:269–77.

Pearre, S., Jr. 1973. Vertical migration and feeding in *Sagitta elegans* Verrill. *Ecology* 54:300–14.

Pejler, B. 1975. On long-term stability of zooplankton composition. *Rept. Inst. Freshwater Res.*, Drottningholm, 54:107–17.

Pennak, R. W. 1953. *Fresh-water invertebrates of the United States*. New York: Ronald Press.

Pope, G. F., and Carter, J. C. H. 1975. Crustacean plankton communities of the Metamek River system and their variation with predation. *J. Fish. Res. Bd.*, Canada, 32:2530–35.

Pope, G. F., Carter, J. C. H., and Power, G. 1973. The influence of fish on the distribution of *Chaoborus* spp. (Diptera) and a density of larvae in the Matamek River System, Quebec. *Trans. Amer. Fish. Soc.* 102:707–14.

Popham, E. J. 1941. The variation in the colour of certain species of *Arctocorisa* (Hemiptera: Corixidae) and its significance. *Proc. Zool. Soc.*, London, *Ser. A* 111:135–72.

———. 1942. Further experimental studies on the selective action of predators. *Proc. Zool. Soc.*, London, *Ser. A* 112:105–17.

Porter, K. G. 1976. The plant-animal interface in freshwater ecosystems. *Amer. Sci.* 65:159–70.

Pourriot, R. 1964. Étude expérimentale de variations morphologiques chez certaines espèces de rotifères. *Bull. Soc. Zool.*, France, 89:555–61.

———. 1974. Relations prédateur proie chez les rotifères: influence du prédateur (*Asplanchna brightwelli*) sur la morphologie de la proie (*Brachionus bidentata*). *Ann. Hydrobiol.* 5:43–55.

Protasov, V. R. 1968. Vision and near orientation of fish. *Israel Program Scient. Trans.*, Jerusalem (1970).

Ratzlaff, W. 1974. Swarming in *Moina affinis*. *Limnol. and Oceanog.* 19:993–95.

Reeve, M. R. 1970. The biology of Chaetognatha. I. Quantitative aspects of growth and egg production in *Sagitta hispida*. In J. H. Steel, ed., *Marine food chains*. Berkeley: University of California Press, pp. 168–89.

Reeve, M. R., Walter, M. A. and Ikeda, T. 1978. Laboratory studies of ingestion and food utilization in lobate and tentaculate ctenophores. *Limnol. and Oceanog.* 23:740–51.

Reif, C. B., and Tappa, D. W. 1966. Selective predation: smelt and cladocerans in Harveys Lake. *Limnol. and Oceanog.* 11:437–38.

Richards, R. C., Goldman, C. R., Frantz, T. C., and Wickwire, R. 1975. Where have all the *Daphnia* gone? The decline of a major cladoceran in Lake Tahoe, California-Nevada. *Verh. Internat. Verein. Limnol.* 19:835–42.

Roberts. T. R. 1972. Ecology of fishes in the Amazon and Congo basins. *Bull. Mus. Comp. Zool. Harvard* 143:117–47.

Rojas de Mendiola, B. 1971. Some observations on the feeding of the Peruvian anchoveta *Engraulis ringen* J. in two regions of the Peruvian coast. In J. D. Costlow, Jr., ed., *Fertility of the sea*, pp. 417–40. New York: Gordon and Breach.

Rosenthal, H. 1972. Uber die Geschwindigkeit der sprungbewegungen bei *Cyclops strenuus* (Copepoda). *Int. Rev. ges. Hydrobiol.* 57:157–67.

Rosenthal, H., and Hempel, G. 1970. Experimental studies in feeding and food requirements of herring larvae (*Clupea harengus* L.). In J. H. Steel ed., *Marine food chains*. Berkeley: University of California Press, pp. 344–64.

Roth, J. C. 1971. The food of *Chaoborus*, a plankton predator in a southern Michigan lake. Ph.D. thesis, University of Michigan.

Ruiter, L. de 1952. Some experiments on the camouflage of stick caterpillars. *Behaviour* 4:222–32.

Schoener, T. W. 1969. Models of optimal size for solitary predators. *Amer. Nat.* 103:277–313.

———. 1971. Theory of feeding strategies. *Ann. Rev. Ecol. Syst.* 2:369–404.

Schultz, D. C., and Northcote, T. G. 1972. An experimental study of feeding behavior and interaction of coastal cutthroat trout (*Salmo clarkii clarkii*) and Dolly Varden (*Salvelinus mama*). *J. Fish. Res. Bd.*, Canada, 29:555–65.

Sebestyén, O. 1931. Contribution to the biology and morphology of *Leptodora kindtii* Focke (Crustacea, Cladocera). *Arb. Ung. Biol. Forschungsi Inst.* 4:151–70.

———. 1960. On the food niche of *Leptodora kindtii* Focke (Crustacea, Cladocera) in the open water communities of Lake Balaton. *Int. Rev. ges. Hydrobiol.* 45:277–82.

Seghers, B. H. 1974. Role of gill rakers in size-selective predation by lake whitefish, *Coregonus clupeaformis* (Mitchill). *Verh. Internat. Verein. Limnol.* 19:2401–05.

Sexton, O. J., and Bizer, J. R. 1978. Life history patterns of *Ambystoma tigrinum* in montane Colorado. *Amer. Mid. Nat.* 99:101–18.

Shapiro, J., Lamarra, V., and Lynch, M. 1975. Biomanipulation: an ecosystem approach to lake restoration. In *Water Qual. Manage. through Biol. Control Symp.*, University of Florida, Gainesville, pp. 85–96.

Simenstad, C. A., Estes, J. A. and Kenyon, K. W. 1978. Aleuts, sea otters, and alternate stable-state communities. *Science* 200:403–11.

Singarajah, K. V. 1969. Escape reactions of zooplankton: The avoidance of a pursuing siphon tube. *J. Exp. Mar. Biol. Ecol.* 3:171–78.

———. 1975. Escape reactions of zooplankton: effects of light and turbulence. *J. Biol. Assn.*, U.K., 55:627–39.

Slobodkin, L. B. 1961. *Growth and regulation of animal populations*. New York: Holt, Rinehart and Winston.

Smith, F. E. 1963. Population dynamics in *Daphnia magna* and a new model for population growth. *Ecology* 44:651–63.

———. 1972. Spatial heterogeneity, stability, and diversity in ecosystems. In E. S. Deevey, Ed., Growth by intussuseption, *Trans. Conn. Acad. Arts and Sci.*, pp. 309–35.

Smyly, W. J. P. 1970. Observations on rate of development, longevity, and fecundity of *Acanthocyclops viridis* (Jurine) (Copepoda, Cyclopoida) in relation to type of prey. *Crustaceana* 18:21–36.

Sprules, W. G. 1972. Effects of size-selective predation and food competition on high altitude zooplankton communities. *Ecology* 53(3):375–86.

———. 1974. The adaptive significance of paedogenesis in North American species of *Ambystoma* (Amphibia: Caudata): an hypothesis. *Can. J. Zool.* 52:393–400.

Stavn, R. H. 1971. The horizontal-vertical distribution hypothesis: Langmuir circulations and *Daphnia* distributions. *Limnol. and Oceanog.* 16:453–66.

Steel, J. H. 1974. *The structure of marine ecosystems*. Cambridge: Harvard University Press.

Stein, R. A. 1976. Sexual dimorphism in crayfish chelae: functional significance related to reproductive activities. *Can. J. Zool.* 54:220–27.

———. 1977. Selective predation, optimal foraging, and the predator-prey interactions between fish and crayfish. *Ecology* 58:1237–53.

Stein, R. A., and Magnuson, J. J. 1976. Behavioral response of crayfish to a fish predator. *Ecology* 57:751–61.

Stein, R. A., Kitchell, J. F., and Kneževič, B. 1975. Selective predation by carp (*Cyprinus carpio* L.) on benthic molluscs in Skadar Lake, Yugoslavia. *J. Fish. Biol.* 7:391–99.

Stein, R. A., Murphy, M. L. and Magnuson, J. J. 1977. External morphological changes associated with sexual maturity in the crayfish (*Orconectes propinquus*). *Amer. Mid. Nat.* 97:495–502.

Stenson, J. A. E. 1974. On predation and *Holopedium gibberum* (Zaddach) distribution. *Limnol. and Oceanog.* 18:1005–1010.

———. 1976. Significance of predator influence on composition of *Bosmina* spp. populations. *Limnol. and Oceanog.* 21:814–22.

———. In press. Predation pressure from fish on two *Chaoborus* species as related to their visibility.

Strickler, J. R. 1975. Swimming of planktonic *Cyclops* species (Copepoda, Crustacea): pattern, movements and their control. In T. Y.-T. Wu, C. J. Brokaw, C. Brennel, eds., *Swimming and flying in nature*. New York: Plenum Press.

———. 1975. Intra- and interspecific information flow among planktonic copepods: receptors. *Verh. Internat. Verein. Limnol.* 19:2951–58.

———. 1977. Observation of swimming performances of planktonic copepods. *Limnol. and Oceanog.* 22:165–70.

Strickler, J. R., and Bal, A. K. 1973. Setae of the first antennae of the copepod *Cyclops scutifer* (Sars): their structure and importance. *Proc. Nat. Acad. Sci.* U.S., 70:2656–59.

Strickler, J. R., and Twombly, S. 1975. Reynolds, diapause, and predatory copepods. *Verh. Internat. Verein. Limnol.* 19:2943–50.

Stromenger-Klekowska, Z. 1960. Cycles annuels des cladocères dans les étangs à poissons. *Int. Rev. ges. Hydrobiol.* 45:215–76.

Strong, D. R. 1972. Life history variation among populations of an amphipod (*Hyalella azteca*). *Ecology* 53:1103–11.

Suffern, J. S. 1973. Experimental analysis of predation in a freshwater system. Ph.D. thesis, Yale University.

Sutherland, J. P. 1974. Multiple stable points in natural communities. *Amer. Nat.* 108:859–73.

Swift, M. C., and Fedorenko, A. Y. 1973. A rapid method for the analysis

of the crop contents of *Chaoborus* larvae. *Limnol. and Oceanog.* 18:795–98.
———. 1975. Some aspects of prey capture by *Chaoborus* larvae. *Limnol. and Oceanog.* 20:418–25.
Szlauer, L. 1965. The refuge ability of plankton animals before plankton eating animals. *Pol. Arch. Hydrobiol.* 13(26):89–95.
Tappa, D. W. 1965. The dynamics of the association of six limnetic species of *Daphnia* in Aziscoos Lake, Maine. *Ecol. Monogr.* 35:395–423.
Taylor, B. E., In press. Life history and the response of zooplankton populations to size-selective predation. Zooplankton Ecology. Symp. Amer. Soc. Limnol. and Oceanog.
Tiegs, O. W., and Manton, S. M. 1957. The evolution of the Arthropoda. *Biol. Rev.* 33:255–337.
Tinbergen, L. 1960. The dynamics of insect and bird populations in pine woods. *Arch. Néerl. Zool.* 13:259–473.
———. 1960. The natural control of insects in pinewoods. I. Factors influencing the intensity of predation by songbirds. *Arch. Néerl. Zool.* 13:265–343.
Uexkull, J., and Kriszat, G. 1934. *Streifzuge durch die Umwelten von Tieren and Menschen*. Berlin: Springer-Verlag.
Vandermeer, J. H. 1970. The community matrix and the number of species in a community. *Amer. Nat.* 104:73–83.
Vermeij, G. J. 1974. Marine faunal dominance and molluscan shell form. *Evolution* 28:656–64.
———. 1976. Interoceanic differences in vulnerability of shelled prey to crab predation. *Nature* 260:135–36.
———. 1977*a*. Patterns in crab claw size: the geography of crushing. *Syst. Zool.* 26:138–51.
———. 1977*b*. The Mesozoic marine revolution: evidence from gastropods, predators, and grazers. *Paleobiology* 3:245–58.
———. 1978. *Biogeography and Adaptation*. Cambridge: Harvard University Press.
Vermeij, G. J., and Covich, A. P. 1978. Coevolution of freshwater gastropods and their predators. *Amer. Nat.* 112:833–43.
Verrier, M.-L. 1948. La vision de vertébrés et les théories de la vision. *Ann. Biol.*, Paris, *Ser. 3,* 24:209–39.
Vinogradov, M. E. 1962. Feeding of the deep-sea zooplankton. *Rapp. Proc. Verb. Cons. Perm. Int. Expl. Mer.* 153:114–120.
Vinyard, G. L., and O'Brien, W. J. 1975. Dorsal light response as an index

of prey preference in bluegill (*Lepomis macrochirus*). *J. Fish. Res. Bd.*, Canada, 32:1860–63.

Virnstein, R. W. 1977. The importance of predation by crabs and fishes on benthic infauna in Chesapeake Bay. *Ecology* 58:1199–1217.

Vlymen, W. J. 1970. Energy expenditure of swimming copepods. *Limnol. and Oceanog.* 15:348–56.

von Ende, C. N. 1979. Fish predation, interspecific predation, and the distribution of two *Chaoborus* species. *Ecology* 60:119–28.

Voris, H. K., and Bacon, J. P., Jr. 1966. Differential predation on tadpoles. *Copeia* 1966:594–98.

Wagler, E. 1941. Die Lachsartigen (Salmonidae). II. Teil. Coregonen. *Handbuch der Binnenfischerei Mitteleuropas III* (6):371–501.

Ware, D. M. 1971. Predation by rainbow trout (*Salmo gairdneri*): the effect of experience. *J. Fish. Res. Bd.*, Canada, 28:1847–52.

———. 1972. Predation by rainbow trout (*Salmo gairdneri*): the influence of hunger, prey density, and prey size. *J. Fish. Res. Bd.*, Canada, 29:1193–1201.

———. 1973. Risk of epibenthic prey to predation by rainbow trout (*Salmo gairdneri*). *J. Fish. Res. Bd.*, Canada, 30:787–97.

Warshaw, S. J. 1972. Effects of alewives (*Alosa pseudoharengus*) on the zooplankton of Lake Wononskopomuc, Connecticut. *Limnol. and Oceanog.* 17:816–25.

Wassersug, R. J. 1975. The adaptive significance of the tadpole stage with comments on the maintenance of complex life cycles in anurans. *Amer. Zool.* 15:405–17.

Wawrik, F. 1966. Zur Kenntnis alpiner Hochgebirgs-Kleingewässer. *Verh. Internat. Verein. Limnol.* 16:543–53.

Weglenska, T. 1971. The influence of various concentrations of natural food on the development, fecundity, and production of planktonic crustacean filtrators. *Ekol. Pol.* 19:428–71.

Wells, L. 1970. Effects of alewife predation on zooplankton populations in Lake Michigan. *Limnol. and Oceanog.* 15:556–65.

Wells, L., and Beeton, A. M. 1963. Food of the bloater, *Coregonus hoyi*, in Lake Michigan. *Trans. Amer. Fish. Soc.* 92:245–55.

Werner, E. E. 1974. The fish size, prey size, handling time relation in several sunfishes, and some implications. *J. Fish. Res. Bd.*, Canada, 31:1531–36.

———. 1977. Species packing and niche complementarity in three sunfishes. *Amer. Nat.* 111:553–78.

Werner, E. E., and Hall, D. J. 1974. Optimal foraging and the size selection of prey by the bluegill sunfish (*Lepomis macrochirus*). *Ecology* 55:1042–52.

———. 1976. Niche shifts in sunfishes: experimental evidence and significance. *Science* 191:404–06.

———. 1977. Competition and habitat shift in two sunfishes (Centrarchidae). *Ecology* 58:869–76.

Werner, E. E., Hall, D. J., Laughlin, D. R., Wagner, D. J., Wilsmann, L. A., and Funk, F. C. 1977. Habitat partitioning in a freshwater fish community. *J. Fish. Res. Bd.*, Canada, 34:360–70.

Werner, R. G. 1969. Ecology of limnetic bluegill (*Lepomis macrochirus*) fry in Crane Lake, Indiana. *Amer. Mid. Nat.* 81:164–81.

Wesenberg-Lund, C. 1939. Biologie der Süsswassertiere: Wirbellose Tiere. Vienna: Springer.

White, G. E., Fabris, G. and Hartland-Rowe, R. 1969. The method of prey capture by *Brachinecta gigas* Lynch 1937 (Anostraca). *Crustaceana* 16:158–60.

Whittaker, R. H., Levin, S. A., and Root, R. B. 1973. Niche, habitat, and ecotope. *Amer. Nat.* 107:321–38.

Wickstead, J. H. 1962. Food and feeding in pelagic copepods. *Proc. Zool. Soc.*, London, 139:545–55.

Wiens, J. A. 1976. Population responses to patchy environments. *Ann. Rev. Ecol. Syst.* 7:81–120.

Wilbur, H. M. 1972. Competition, predation, and the structure of the *Ambystoma-Rana sylvatica* community. *Ecology* 53:3–21.

Williams, G. C. 1975. Sex and evolution. *Princeton Monogr.* Princeton, N. J.: Princeton University Press.

Wilson, D. S. 1973. Food-size selection among copepods. *Ecology* 54:909–14.

Wilson, M. S. 1959. Calanoida. In W. T. Edmondson, ed., *Freshwater biology.* New York: Wiley, pp. 738–94.

Wong, B., and Ward, F. J. 1972. Size selection of *Daphnia pulicaria* by yellow perch (*Perca flavescens*) fry in West Blue Lake, Manitoba. *J. Fish. Res. Bd.*, Canada, 29:1761–64.

Wright, J. C. 1965. The population dynamics and production of *Daphnia* in Canyon Ferry Reservoir, Montana. *Limnol. and Oceanog.* 10:583–90.

Wynne-Edwards, V. C. 1962. Animal dispersion in relation to social behaviour. New York: Hafner Publishing Co.

Zaret, R. E. In press. The animal and its viscous environment. In Zooplankton Ecology. *Symp. Amer. Soc. Limnol. and Oceanog.*

Zaret, T. M. 1969. Predation-balanced polymorphism of *Ceriodaphnia cornuta Sars*. *Limnol. and Oceanog.* 14:301–03.

———. 1971. The distribution, diet, and feeding habits of the atherinid fish *Melaniris chagresi* in Gatun Lake, Panama Canal Zone. *Copeia* 1971:341–43.

———. 1972*a*. Predator-prey interaction in a tropical lacustrine ecosystem. *Ecology* 53:248–57.

———. 1972*b*. Predators, invisible prey, and the nature of polymorphism in the Cladocera (Class Crustacea). *Limnol. and Oceanog.* 17:171–84.

———. 1975. Strategies for existence of zooplankton prey in homogeneous environments. *Verh. Internat. Verein. Limnol.* 19:1484–89.

———. 1979. Predation in freshwater fish communities. In H. Clepper, ed., *Predator-prey systems in fisheries management.* Washington, D.C.: Sport Fishing Institute, pp. 135–143.

———. In press. The effect of prey motion on planktivore choice. Zooplankton Ecology. *Symp. Amer. Soc. Limnol. and Oceanog.*

Zaret, T. M., and Kerfoot, W. C. 1975. Fish predation on *Bosmina longirostris*: body-size selection versus visibility selection. *Ecology* 56:232–37.

Zaret, T. M., and Paine, R. T. 1973. Species introduction in a tropical lake. *Science* 182:449–55.

Zaret, T. M., and Rand, A. S. 1971. Competition in tropical stream fishes: support for the competitive exclusion principle. *Ecology* 52(2):336–42.

Zaret, T. M., and Suffern, J. S. 1976. Vertical migration in zooplankton as a predator avoidance mechanism. *Limnol. and Oceanog.* 21:804–13.

Zipser, E., and Vermeij, G. J. 1978. Crushing behavior of tropical and temperate crabs. *J. Exp. Mar. Biol. Ecol.* 31:155–72.

Index